Breeding for Crop Improvement

About the Author

Venkata R Prakash Reddy received his B.Sc (Ag) degree from Acharya N G Ranga Agricultural University, Hyderabad, Andhra Pradesh and M.Sc (Ag) from Department of Genetics and Plant Breeding, University of Agricultural Sciences, Dharwad, Karnataka. Where he received University Gold medal. He is a recipient of Inspire Fellowship of Department of Science and Technology for PhD programme. Currently working as Assistant Professor at Acharya N G Ranga Agricultural University, Agricultural College, Mahanandi and he has published several research, review papers in various national and international journals. He has already authored three books titled Heterosis and Combining ability studies in Sunflower and Key notes on Genetics and Plant Breeding and Key notes on Plant Biotechnology. His research interests include Plant Breeding and Molecular Biology.

Breeding for Crop Improvement

Venkata R Prakash Reddy
Assistant Professor
Dept. of Genetics and Plant Breeding
Acharya N G Ranga Agricultural University
Agricultural College
Mahanandi-518502

SSPH

SATISH SERIAL PUBLISHING HOUSE
403, Express Tower, Commercial Complex
Azadpur, Delhi-110033 (India)
Phone : 011-27672852, Fax : 91-11-27672046
e-mail : info@satishserial.com, hkjain1975@yahoo.com
Website : www.satishserial.com

Published by :

SATISH SERIAL PUBLISHING HOUSE
403, Express Tower, Commercial Complex, Azadpur, Delhi-110033 (INDIA)
Phone : 011-27672852 Fax : 91-11-27672046
E-mail : info@satishserial.com, hkjain1975@yahoo.com

Reprint 2020

ISBN 978-93-86200-20-4

Composed, Designed & Printed in India

Prof. E.A. Siddiq
FNA, FNASc., FNAAS,
Sertior Scientist, NASI Platinum Jubilee Fellowship
Hon, Professor (Biotechnology) Professor Jayashankar
Telangana State Agricultural University
Adjunct Faculty, Centre for DNA Fingerprinting and Diagnostics
Former National Professor (ICAR) & Deputy Director General (CS) ICAR

FOREWORD

Plant breeding is the art of managing genetic variability keeping in view the changing varietal demand of stakeholders. Crop improvement is accomplished employing many different breeding techniques ranging from selecting plants of interest in the native germplasm knowing by "pureline selection" to more complex inter-varietal hybridization followed by selection using molecular tools. Classical plant breeding uses deliberate interbreeding (crossing) among closely or distantly related but cross compatible individuals to evolve new crop varieties with desired trait combinations. Reproductive system that allows recombination is the major source of variability that is scientifically managed to achieve breeding goals. In flowering plants, a highly diverse array of floral traits determine the reproductive system, varying from obligated cross fertilization to obligated self-fertilization. Selfed progeny of crosses offer reproductive assurance and two-fold transmission of genes than crossed progeny, which in turn affects the choice of breeding methods used for improvement of self and cross pollinated crops.

It is the science of crop improvement that plays important role in progressively improving crop plants in desired directions. Conscious efforts to improve crop plants of interest to mankind date back to the period of domestication and settled agriculture. Scientific crop improvement started following the rediscovery, elucidation and application of Mendel's Law of inheritance over one hundred years back. It was such knowledge based crop breeding that led to development of major landmark varieties/hybrids across crop plants, exploiting which the world could meet the continuously rising demand projections of food, feed and fiber over the last century. They include broadly plant type based high yielding dwarf varieties of wheat and rice, hybrid technology in both cross and self-pollinated crops like maize, sorghum, pearlmillet, rice cotton and a long list of vegetable crops. As a result many countries including India could become self-sufficient in their need of food and other essential commodities.

Inspite of what has been achieved the challenge of meeting the rising the needs of the growing global population is still haunting us. By 2050, our food grain need, for instance, is estimated to be 70-80% more than what we produce now. With fast shrinking natural resources like arable land area and water, meeting the demand is going to be a major challenge for us today. This warrants the need of innovative breeding strategies and new genetic variability to evolve varieties that would help find solution to such constraints and ensure achievement of the future production targets.

Keeping in view the foregoing, this book entitled "Breeding for Crop Improvement" has been meticulously planned and designed giving due emphasis to new breeding/selection strategies underway in crop improvement research. Placing emphasis on major cereals, millets, oilseed and pulse crops, cotton and vegetable crops. 1 feel that a comprehensive book like this would be of great value to young researchers and students of Agriculture. Also, it has been written keeping in view latest syllabi, as prescribed by the 5" Deans' Committee. Indian Council of Agricultural Research. 1 compliment the sincere efforts of Sri. V.R. Prakash Reddy, Assistant Professor. Department of Genetics and Plant Breeding, ANGRAU who has taken pain to bring out this book. 1 hope and believe that this book would be very useful and instructive for under graduate and post-graduate students in all Agricultural Universities in the country.

(E.A. SIDD1Q)

Institute of Biotechnology
Professor Jayasnankar Teiangana State Agricultural University
Rajendranagar, Hyderabad-500 030
Res.: 81, Happy Homes Colony, Upperpally, Hyderabad-500 048
Ph.: 24018625 (R) 09032326727 (C)
E-mail : easiddiq@gmail.com / easiddiq@rediffmail.com

ACHARYA N G RANGA AGRICULTURAL UNIVERSITY GUNTUR, ANDHRA PRADESH

Dr. B. Govinda Rao
Professor and University Head
Dept of Genetics and Plant Breeding
ANGRAU

PREFACE

Human being is almost dependent on plants for his food and all other needs. Plants are major source directly or indirectly of most food, fodder clothing, fuel, drugs etc. Considering the prime importance of plants, human being have long been concerned with developing plant types better suited to satisfying their needs. Plant breeding, also known as the Science of Crop Improvement or Crop Improvement technology, plays an important role in improving crop plants in relation to their economic use for mankind. Plant breeding has been practiced for thousands of years for developing new varieties that are high yielding, resistant to biotic and abiotic stresses and better adaptation to different environmental conditions. It is practiced worldwide by individuals such as gardeners and farmers, or by professional plant breeders employed by organizations such as government institutions, universities, crop-specific industry associations or research centers. It resulted in "Green Revolution" which increased agricultural production worldwide, particularly in the developing world most markedly in the late 1960s.

At present future of plant breeding looks more promising. Modern plant breeding is the science of improving plants to achieve these needs and better fit production environments, but is a long-term proposition. Keeping in view of these aspects, this book entitled "Breeding for crop improvement" has been designed to provide theoretical and practical knowledge on selections and hybridization principles in crop improvement, breeding methods in self and cross pollinated crops, breeding for biotic and abiotic stress resistance, breeding in cereals, millets, oilseeds, fibre crops, vegetable crops, sugar crops, narcotics etc. Particularly, this book will be useful for understanding floral biology, anthesis and pollination, breeding objectives and breeding methods of different crop plants for successful crop improvement. I complement the sincere efforts of Sri. V.R. Prakash Reddy, Assistant Professor, Department of Genetics and Plant Breeding, ANGRAU to bring out this book by considering latest syllabi, as prescribed by the V Deans' Committee recommendations appointed by the Indian Council of Agricultural Research. I hope this text book will be very useful and instructive for the under graduates, post-graduates and doctoral students of all Agricultural Universities.

(Dr. B. Govinda Rao)

CONTENTS

Chapter-1

Plant Breeding: Selection and Hybridization Principles

Plant breeding is concerned with principles and methods required for favorably changing the genetic constitution of crop plants in terms of agronomically as well as economically. Thus plant breeding can be defined "as an art and science" of improving the genetic makeup of plants in relation to their economic use for the mankind. Recently plant breeding has been described as a technology of developing superior crop plants for various purposes (Riley, 1978). Plant breeding is also considered as the current phase of crop evolution (Simmonds, 1979). Since plant breeding is a deliberate process, it is destined for definite goals to bring about improvement in crops in respect of productive capacity, quality consideration, resistance to biotic and abiotic stresses, adaptive ability and suitability to new crop husbandry etc. These well defined breeding objectives must be achieved through a priori knowledge of the factors governing the success of plant breeding procedures such as the nature of gene action commanding phenotypic expression, genetic diversity showing range of heritable variation and genotype with environment interaction influencing phenotypic stability etc. Thus plant breeding mainly involves the systematic production of crop populations exhibiting genetic segregation/ variation and selection within that population to establish lines with favorable allele combinations.

Production of variation is achieved through sexual crossing particularly of cultivated lines. Two parents who have expression of the desirable characters between them are intercrossed and the subsequent generations examined for plants with the desired characters in new combinations, i.e. look for recombinants. Clearly the choice of parents is critical. Induction of variation is also arising from the occasional and rare mistakes called mutations that occur in the otherwise faithful replication of the DNA in all organisms. The breeder generally uses the natural variation that already exists within the species. For virtually all characters we only need to look or measure any character to observe variation in their expression (phenotype), and often this reflects not just variation produced by differences in the environment in which the plant happens to be growing, but also genetic variation that is heritable (genotype). This naturally occurring source of heritable variation accounts for most of the responses that have been made in plant breeding. The relation between phenotype and genotype can be written as

Phenotype = Genotype + Environment + (Genotype × Environment)

Thus, the less that the environment affects the character either directly or by interaction of the genotype that will be gained by simply observing the phenotype.

Plant breeding: Selection

Selection is basic to any crop improvement. The term selection coined by Lerner and it is non-random differential reproduction of genotype or gradual heritable changes in the genetic makeup of genotypes without mutation or hybridization. Simply isolation of desirable plant types from the population. It is one of the two fundamental steps of any breeding programme viz., creation of variation and Selection. The effectiveness of selection primarily depends upon the degree to which phenotype reflects the genotype. There are two agencies involved in carrying out selection: one is Nature itself (Natural selection) and the other is man artificial selection.

The main function of the selection is to discriminate between individuals. This is possible only when sufficient variation is present in the material subjected to selection pressure. Thus, selection acts on the existing variation it cannot create new variation. Only the selected individuals are permitted to contribute to the next generation / progenies. Hence the variation among individuals to be selected must be genetic in nature, since it is the genetic variation that tends to close the gap between phenotype and genotype. Environmental variability cannot be of any use under selection.

Elements of selection: Selection does not act on the genes directly but acts on the genotypes by altering their relative contribution to the progeny as a result of which frequency is changed under the influence of selection. Changes in gene frequencies at the loci conditioning the trait selected are observed by shift in the population mean. Since gene frequency alteration cannot be traced individually for each locus involved, but change of population mean can be readily traced to examine the rate at which gene/allelic frequencies are changed under selection pressure. The main determinants or elements of selection process for a quantitative trait includes Selection intensity, Selection differential, Heritability and Selection response

Selection intensity (i): It refers to amount of selection applied expressed as proportion of population favored and it is inversely proportional to percentage proportion selected. High selection intensity (large 'i') means stringent selection pressure. In real selection experiment, the desired alleles become preponderant after each cycle of selection so selection intensity 'i' also change. Low selection pressure (small 'i') may cause large populations to be handled that unnecessarily waste the time and resources. Large 'i' i.e. high selection pressure that leads to allelic erosion due to genetic drift. The selection intensity 'i' equals to 2.06 to 1.76 (i.e. around 5-10 per cent of individuals selected) has generally been found appropriate in plant populations. Under natural selection intensity is expressed as relative number of offspring produced by different genotype and known as selection co-efficient (s) = i/σ_p. Where, i = intensity of selection, s = selection differential and σ_p = phenotypic standard deviation.

Selection differential (s): It is average superiority of selected individuals over the mean of population of their origin in same parental generation before selection is made. The wider the phenotypic variability, greater the selection differential's'. The standardized selection differential i.e., s/σ_p is also known as intensity of selection. So, $S=i\sigma_p$. It means selection differential's' is directly proportional to selection intensity 'i' and phenotypic standard deviation 'σ'_p.

Heritability (H): It refers to the measure of heritable portion of observed variability or measure of dependence of breeding value on phenotypic value or regression of breeding value on phenotypic value i.e. it is the ratio of genetic variance to total variance.

Broad sense heritability (H_{bs}) = $V_g/V_p = V_g/(V_g+V_e)$

Heritability estimated from the above formula is known as the broad sense heritability. This is valid when homozygous lines are studied. But when segregating generations are studied genotypic variance consists of additive variance, dominance variance and variance due to epistasis. Dominance variance is important when we are dealing with hybrids i.e. F_1 generations. In self pollinated crops we release varieties only after making them homozygous lines. Hence additive variance is more important in such cases. The proportion of additive genetic variance to the total variance is known as narrow sence heritability. If heritability is very high for any character it can be improved. Improvement of characters with low heritability is very difficult.

Narrow sense heritability $(H_{ns}) = V_A/V_P$

Selection response or genetic advance under selection (R)

It is deviation of the progeny mean (x) from the original (unselected) population mean (ì). It is also known as genetic advance or genetic gain under selection.

$R = X - \mu$.

The basic idea of selection response stems from the fact that when a population with considerable genetic variability is subjected to selection, change in gene frequency is obvious and resulted in shift in population mean. The shift in population mean reflected in next generation in form of selection advance.

Response to selection is mainly governed by three factors viz., high heritability (h^2), high selection intensity (i) and large selection differential. Variability is important here because high selection differential 's' results from from high selection intensity 'i', but both selection intensity and selection differential are dependent upon standard deviation (σ_p) $s = i\sigma_p$

$R = h^2 s = h^2 \sigma_p$

Selection in self and cross pollinated crops: Selection operates on genetic variability, which is of diverse nature in self and cross pollinated crops with respect to magnitude and type. Thus, principles involved in the selection for some traits in self pollinated crops are a little different from those in cross pollinated crops though the ultimate goal is same i.e. genetic advance.

In **self pollinated crops**, as a consequence of regular inbreeding, free variability floats between families and is distributed among a large number of homozygous genotypes. Therefore, a single cycle of selection is enough to freeze the free variability i.e. fixation of genotypes is achieved with a selection in self pollinated crops. Further response to selection is possible only when raw variability is generated by mutation / hybridization between superior single plants followed by selection in segregating generations. In F_2 generation, the selection is generally practiced at low selection intensity and the proportion saved for F_3 is large. This is so because the partitioning of variation into genetic and non-genetic causes, particularly for traits with low heritability. Hence the selection at this stage is unreliable. Based on this principle, the simply inherited traits and traits with high heritability may receive greater attention in selection during early segregating generations (F_2 and F_3). Conversely, complex traits with low heritability should be tackled in later generations when considerable homozygosity has been achieved. This is the reason mainly for selection within families is practiced in early segregating generations and between families in advanced generations in self pollinated crops. This is the basic principle involved in the selection practices in almost all self pollinated crops.

In **cross pollinated crops**, as a consequence of reservoir of variability always available for selection, and the variability lost under selection is promptly replenished by free gene flow (because of open pollination) among constituent members of the population. However, in cross pollinated crops, individual genotypes are heterozygous. So, the aim of selection is to shift the gene frequency at a stable equilibrium based on the Hardy- Weinberg principle. Because of nature and magnitude of variability in cross pollinated crop, selection becomes a slow process. Response to selection in cross pollinated crops can be classified in to five main types based on the availability of free variability as well as potential variability (Allard, 1960) This includes

a. Rapid initial gain under selection

b. Slow, steady and long continued response

c. Slow response culminating a plateu

d. Little or no response to selection

e. Selection response after plateuing

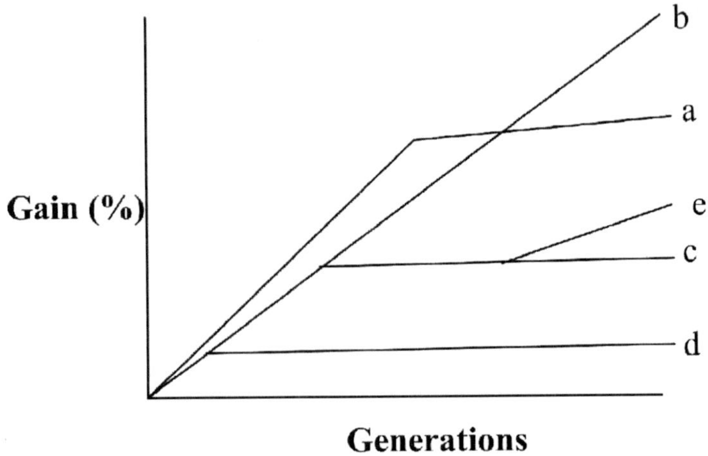

Patterns of selection response (a-e) for quantitative trait

Rapid initial gain under selection; it is mainly applicable for adjustment of quasi-quantitative traits (simply inherited but not necessarily monogenic). E.g. Resistance, pigmentation and leaf angle etc. The cause for rapid gain underlies the increase in the frequency of major gene, which is expected to be large until these genes are fixed by selection. Subsequently, slow progress due to slow increase in minor gene frequency coupled with further small changes in the frequency of major genes. And also, here the genetic variability is reduced due to the fixation of major genes.

Slow, study-long continued response; it is apllicable for quantitative traits is not rewarded as well as for the quasi-quantitative trait. In this we can observe the gradual shift in the population mean, followed by little shrinkage of genetic variability. Selection response mainly due to prolonged availability of variability over generations. Selection does not seem to increase or decrease variability very drastically.

Slow response culminating a plateau; Reasons for slow response is mainly due to lack of adequate additive variability (σ^2_A) and/or faulty estimation of heritability (h^2) and genetic advance. And also unfortunate choice of breeding strategy may also contribute to such a pattern.

Little or no response to selection; Directional selections for complex characters like grain yield leads to practically little or no response to selection. Reasons for no or little response are due to lack of genetic variability and limited exploitation of additive variability (σ^2_A) or low heritability (h^2) for quantitative trait like yield.

Selection response after plateuing; this situation represents a typical behavior of mendalian population with respect to storage and release of variability. The linked loci carry variability in a population in two forms. Free variability: If free variability is available, response to selection will be reflected in the actual initial gain. After it is exhausted, response will assume plateuing. This period continues till the Potential variability is converted in to free variability by segregation. This follows a second flush of response to selection.

Asymmetrical response to selection: It refers to unequal response to selection in the opposite direction. And it is characteristic feature of two way selection and disruptive selection. The cause for this response is due to the loci contributing negatively to the co-variance and having $p \neq q$ (here, p and q are allelic frequencies of both dominant and recessive alleles) i.e. loci showing gene frequency other than 0.5 i.e. $p \neq q$. In other words, sensitivity of genetic covariance to changed gene frequency brought out by selection tends to contribute to asymmetry of response.

Correlated response to selection: It refers to natural consequence of selection for a character selected, accompanied by change in other related characters, not subjected to selection pressure. The phenomena observed much earlier by Darwin. This is causes for homeostatic regulation and co adaptation of gene pool. Correlated response to selection is the basic property of quantitative traits. Whereas no correlated response in qualitative traits.

Genetic cause of such response explained by following theory

1. Genetic correlation theory on quantitative basis: Characters genetically related to each other tend to move in same direction under selection. There are two mechanisms that cause genetic correlation, hence correlated response. i.e. Pleiotrophy - developmental correlation and linkage
2. Polygenetic inheritance theory on purely qualitative basis. Selection favoring alleles in certain block affect the frequencies of alleles that are linked to them. Association of loci and blocks located on different chromosome will also have similar effect.

Selection for multiple traits: These are three basic strategies to simultaneously select multiple traits: tandem selection, independent culling and selection index.

Tandem selection: In this, the breeder focuses on one trait at a time (serial improvement). One trait is selected for several generations, thus another trait is focused on for the next period. It is effective when genetic correlation does not exist between the traits of interest, or when the relative importance of each trait changes throughout the years.

Independent culling: Also called truncation selection, independent culling entails selecting for multiple traits in one generation. E.g. Traits A, B and C, and breeder may select 50% plants per family for A on phenotypic basis and from the group select 40% plants per family based on trait B; Finally, from that subset, 50% plants per family are selected for trait C, giving a total of 10% selection intensity (0.5 × 0.4 × 0.5).

Selection index: The problem with this approach is that as more traits are selected for, the less the selection pressure that can be exerted on any one trait. In selecting on a multivariate phenotype, the breeder explicitly or implicitly assigns a weighting scheme to each trait, resulting in the creation of a univariate trait (an index) that is then selected. The index is the best linear prediction of an individual's breeding value.

Plant breeding: Hybridization

Hybridization is the process of crossing two genetically different individuals to create new genotypes. The first natural hybridization was recorded by Cotton Mather (1716) in corn. The first artificial interspecific plant hybrid was produced by Thomas Fairchild in 1717. It is commonly known as 'Fairchild mule'. Hybridization was first of all practically utilized in crop improvement by Joseph Koelrauter in 1760. Mendel onward, the hybridization had become the key method of crop improvement. Today, it is the most common method of crop improvement, and the vast majority of crop varieties have resulted from hybridization. The main objectives of the hybridization include; creation of variable population for the selection of types with desired combination of characters, combining the desired characters into a single individual and to exploit and utilize the hybrid varieties.

There are various types of hybridization, which includes intra-varietal hybridization means crosses are made between the plants of the same variety, inter-varietal or intraspecific hybridization means crosses are made between the plants belonging to two different varieties, inter-specific or intergeneric hybridization means crosses are made between two different species of the same genus and introgressive hybridization means transfer of some genes from one species into the genome of the other species. The crosses between different species of the same genus or different genera of the same family are also known as distant hybridization or wide crossing. Such crosses are called distant crosses. It involves the following steps.

1. Selection of parents: The selection of parents should be done with great care, taking into consideration the objectives, so that the chances of selecting the desirable genotypes would be more from the segregating population. As far as possible the parental types (female parent) must be selected from the local types, because they are best adaptable. While selecting the parents all the important characters to combine must be kept in mind.

2. Selfing the parents: This is the second step and is essential for elimination of undesirable characters and also for obtaining the homozygous individuals. Selfing need to be done in self pollinated crops while in cross pollinated crops it has to be performed artificially.

3. Emasculation: This is a very important step in hybridization. Emasculation is the removal of male organs in bisexual flower before the anthers ripe and shed the pollen on the stigma to prevent self fertilization.

4. Bagging: The selected male and female flowers have to be bagged separately to prevent the contamination in staminate flowers and cross pollination in pistillate flowers. Bagging has to be done invariably the precious evening of the day of crossing. Different types of bags are used for this purpose. It is also essential to puncture the paper bags with minute holes to prevent mould formation and also to provide ventilation for the developing seeds.

5. Crossing: The desired pollen/anthers have to be collected from the male parent either in Petri dishes or in paper bags. The collected pollen has to be brushed or dusted on the stigmatic surface of the emasculated flower (female parent). After crossing (hand pollination) the flowers are again bagged.

6. Tagging and labeling: The crossed flowers have to be properly bagged and labeled with the following. Name of parents (female

and male), date of emasculation, number of flowers/spikelets emasculated, date of crossing and number of flowers/spikelets crossed etc.

7. Harvesting: The crossed seeds (F_1) have to be harvested after maturity and preserved after drying to safe moisture level to raise the F_1 generation.

Selfing: In self pollinated crops, artificial selfing is not necessary as they are naturally self fertilized. But in cross pollinated crops, the inflorescence should be enclosed in bags before anthesis to avoid cross pollination. This will prevent the foreign pollen to reach the stigma. Seed set is frequently reduced in ear heads enclosed in bags because of excessive temperature and humidity inside the bags. In the case of wheat, rice, barely, groundnut etc., the plant is permitted to have self pollination and the seeds are harvested. In crops like cotton which have larger flowers the petals may fold down the sexual organs and fasten, there by pollen and pollen carrying insects may be excluded. In certain legumes which are almost insect pollinated, the plants may be caged to prevent the insect pollination. In crops like castor the whole inflorescences enclosed in a bag and in maize selfing are done by collecting the pollen from the tassel and dusting on the silks of the same palnt.

Emasculation: Removal of anthers or stamens or killing the pollen of a flower without the female reproductive organ is known as emasculation. In bisexual flowers, emasculation is essential to prevent of self pollination. In monoecious plants (E.g. Castor and coconut), male flowers are removed or male inflorescence (E.g. Maize) is removed. In species with large flowers (E.g. Cotton) hand emasculation is accurate and it is adequate. The following methods commonly used for emasculation in crop plants.

1. Forced open method: In this method the flowers are opened by force before it opens naturally and the anthers are removed with a forceps. This technique is useful in large size flowers like tomato, brinjal and chillies etc.

2. Clipping method: This technique is usually used in cereals like rice and wheat. In this method top $1/3^{rd}$ portion of the spikelets are cut with scissors and the anthers are removed with forceps.

3. Ring cut method: A circular cut is made at the base of the corolla, 2mm away from the ovary. The corolla is removed and the anthers

are picked out with the help of a forceps. This method is generally used in bhendi, cotton and in leguminous crops.

4. Keel rupturing method: In papilinaceous (fabaceae) flowers such as black gram, green gram, red gram the keel petal is ruptured and the stamens are removed with the help of forceps.

5. Hot water treatment: The removal of stamens with forceps is time taking and tedious in plants having small sized flowers such as rice, jowar, ragi etc, In such cases dipping the panicles in hot water for a definite period of time at a desired temperature is effective. This makes the pollen grains inviable without affecting the stigma and ovary. The panicles can be dipped in a thermo flask containing hot water. The temperature of water and time of dipping differ from crop to crop and variety to variety in rice 45-52^{0}C for 2-10 minutes is found effective.

6. Cold water treatment: Cold treatment like hot water treatment kills the pollen grains without damaging gynoecium. In the case of rice, treatment with cold water 0.6^{o}C kills the pollen grains without affecting the gynoecium. This is less effective than hot water treatment.

7. Suction method: It is useful in species with small flowers. A thin rubber or glass tube attached to a suction hose is used to suck the anthers from the flowers. The amount of suction used is very important which should be sufficient to suck the pollen and anthers but not the gynoecium. In this method considerable self pollination, up to 10% is likely to occur. Washing the stigma with a jet of water may help in reducing self pollination. However self pollination cannot be eliminated in this method.

8. Chilling method: Chilling of plants for a particular length of time to allow the temperature to make the pollen sterile. In wheat, chilling treatment for 13 to 14 hours at 27-30^{0}C would make the pollen sterile.

9. Relative humidity method: The ear heads are given a special treatment so that the relative humidity around the panicles is increased. For this purpose the inflorescence is covered with moist cloth bag. Due to this the relative humidity increases forcing the anthers to come out, without dehiscing. The stamens are then removed with forceps. E.g. Rice.

10. Hot air method: The panicles are enclosed in a brown paper bag one to two hours before anthesis. After half an hour the temperature within the bag raises forcing the anthers to emerge out. The anthers are then removed with the help of forceps.

11. Chemical method: By placing some of the chemicals called 'Gametocides' such as 2,4-D, Maleic hydrazide, NAA, TIBA and Etherel etc. are used to induce the male sterility.
12. Alcohol treatment: The flower buds are dipped in ethyl alcohol for 5 to 10 minutes for making the pollen sterile. E.g. Lucerne floral buds dipped in 57% ethyl alcohol for 10 minutes are male sterile.
13. Genetic emasculation: In crops like maize, jowar, bajra, potato and sunflower male sterility occurs due to cytoplasmic and genetic causes or both. Emasculation is not needed in such cases. And in crops like tobacco, cabbage, cauliflower etc, pollination with the desired pollen is possible without doing emasculation due to the presence of self incompatibility.

Pollination: Pollen grains collected from a desired male parent should be transferred to the emasculated flower. This is normally done in the morning hours during anthesis. The flowers are bagged immediately after artificial crossing. The emasculated flowers can be pollinated in different ways:

1. When the flowers produce small quantity of pollen, it is transferred with the help of camel hair brush. E.g. Mango, chillies, tomato and gingelly.
2. When the flowers produce large amount of pollen, it will be collected in glazed butter paper bags and dusted on the emasculated flowers. E.g. Bajra, jowar and maize
3. In case of clipping method of emasculation, the individual mature anther may be collected and inserted into the emasculated spikelets/ flowers. E.g. Paddy and wheat etc.
4. When the flowers are possessing numerous anthers the individual flower is collected after the corolla and calyx are removed and will directly made in contact with the stigma of the emasculated flower. E.g. Bhendi, guava and pome granate.
5. The inflorescence of male and female parents are enclosed in a single bag and allowed for pollination. Seeds from the female parent are harvested and raised in a nursery. Based on the marker gene character the F_1 seedlings are identified.

Genetic significance of pollination method: Plants that are normally self pollinated differ in genetic makeup from plants that are normally cross pollinated. In self pollinated crops, it is the rule that plants will be homozygous or nearly so. This is because gene pairs homozygous (AA or aa) will remain homozygous with self pollination, gene pairs with heterozygous (Aa) will segregate, producing homozygous and heterozygous progeny in equal proportion with self pollination and heterozygosity is reduced by one-half with each successive self fertilization. This was illustrated below.

No. of generations of selfing	Generation	Frequency % AA	Aa	aa	Frequency % Homozygosity	Heterozygosity
1	F_1	0	100	0	0	100
2	F_2	25	50	25	50	50
3	F_3	37.5	25	37.5	75	25
4	F_4	43.75	12.5	43.75	87.5	12.5
5	F_5	46.87	6.25	46.87	93.73	6.25
6	F_6	48.43	3.12	48.43	96.87	3.12
7	F_7	49.21	1.56	49.21	98.43	1.56
8	F_8	49.60	0.78	49.60	99.21	0.78
9	F_9	49.80	0.39	49.80	99.60	0.39
Formula		$(2^m-1/2^{m+1})$	$(1/2^m)$	$(2^m-1/2^{m+1})$	$(2^m-1/2^m)$	$(1/2^m)$

Each generation of self fertilization increases the homozygosity by 50% and reduces the heterozygosity by 50% i.e. increasing the homozygosity with a corresponding decreasing in heterozygosity. The proportion of completely homozygous plants after 'm' number of selfings with 'n' number of genes segregating is $(2^m-1/2^m)^n$. Thus continued self pollination has two main effects on the population, first all the plants in the population become completely homozygous and second, the population of self pollinated species represents a mixture of several homozygous genotypes. This permits the breeder to develop pure homozygous and homogeneous varieties either by pure line selection or hybridization followed by selection methods like pedigree or bulk methods etc in self pollinated crops.

In cross pollinated crops, each plant receives blend of pollen from large number of individuals with different genetic constitution. So these populations exhibit high degree of heterozygosity and heterogeneity with tremendous free and potential genetic variation which is maintained in steady state by free gene flow among individuals within the population. This permits the breeder to follow two approaches; one is development of hybrids by hybridizing the selected contrasting parents. Another is population improvement, it involves a stepwise elimination of deleterious and less productive alleles through repeated cycles of selective mating of more productive genotypes E.g. Recurrent selections. Population improvement is slow, steady and long term programme whereas production of hybrids is aimed to maximize the genetic gains in much less time.

□□□

Chapter-2

Breeding Methods in Self Pollinated Crops

The basic methods by which new varieties of self pollinated crops originated are introduction, selection and hybridization. As plant breeding procedures developed, the three methods of breeding tended to follow the sequence listed. The initial step was to introduce varieties from numerous sources that were adapted locally. Because many of the early varieties were genetic mixtures, selection was practiced to isolate the desirable type from the mixture. Later, the desirable characteristics of the varieties identified were combined by hybridization. Through hybridization, segregating populations are developed, from which genotypes may be selected with a combination of genes superior to either parent. In practice, breeders may deviate considerably from the methods as outlined, although the principles upon which their procedures are based may be unchanged. In this chapter attention is given to the selection and hybridization procedures in self pollinated crops and to related breeding procedures such as mass selection, pureline selection, pedigree method, bulk method, single seed descent method, back cross method and multiline breeding *etc*. of these methods, mass selection and pure line selections acts only on existing variation but not for create of variation. But other methods follows the creation of variation first and then operates the selection *i.e.* hybridization followed selection methods.

Mass selection

Mass selection is often described as the oldest method of breeding. However, this by no means makes the procedure outdated. As an ancient art, farmers saved seed from desirable plants for planting the next season's crop, a practice that is still common in the agriculture of many countries. This method of selection is applicable to both self and cross pollinated species. In mass selection, selection is mainly based on the phenotypic superiority of plants i.e. large numbers of plants of similar phenotypes are selected and their seeds are mixed to constitute a new variety.

The purpose of mass selection is population improvement through increasing the gene frequencies of desirable genes. But the improvement is limited to the genetic variability that existed in the original population i.e. selection operates on existing variability not create the variability. Mass selection is mainly applied to maintain the purity of an existing cultivar that has become contaminated or is segregating by rouging of the off types or plants with undesirable traits. It is known as *Negative mass selection* and is generally used for varietal purification in seed production and certification programmes. In some cases, breeders follow the selection and advancement of large number of plants that are desirable and uniform for the trait of interest. It is known as *Positive mass selection* and is generally used for improvement of old varieties or land races. The general procedure for mass selection in developing new varieties is outlined below. But the procedure may be modified according to the needs of programme.

Year	Mass selection	Steps
1st	Base Population ↓	From source population select the plants with similar and desirable characters.
2nd	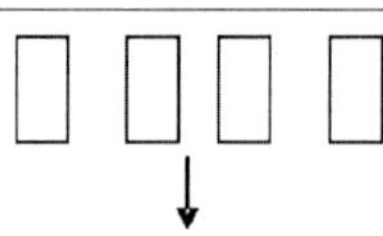	Preliminary yield trials with selected plants along with standard check. And phenotypic characters of the new variety were critically evaluated.
3rd to 5th	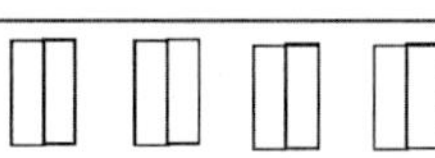	Advanced yield trials with selected plants
6th	Release best variety	Seed multiplication and distribution

Mass selection may be conducted in both self and cross populations, but the genetic consequences will be different. In self pollinated populations, the persistence of inbreeding will alter population gene frequencies by reducing heterozygosity from one generation to the next. In cross pollinated populations, the gene frequencies are expected to remain unchanged unless the selection of plants was biased to change the frequency of alleles that control the trait of interest. Although mass selection is simple, rapid and straight forward method, it mainly suffers from the drawbacks like selection is based on phenotype not on real genotype because there is no control over the environment and selection is also female oriented selection because there is no control over pollination. These defects can be overcome by progeny selection; it is critical tool to differentiate the contribution of genes and environment towards phenotype. Mass selected end product shows considerable genetic variability and are not uniform as like pure line varieties. But it provides better buffering capacity against environmental changes and also suitable for another round of mass selection after few years for improving the variety further.

Pureline selection: Development of new variety through identification and isolation of single best plant progeny is known as pure line selection. Pure line refers to the homogeneous progeny of a self pollinated homozygous plant. Pure line suggests that a cultivar has identical alleles at all loci. Pure lines are generally highly uniform, homogeneous and narrow adapted lines. Genetic basis of pure line explained by Danish biologist Johannsen in 1903 while working with Princess variety of common bean (*Phaseolus vulgaris*). The pure line theory may be summarized as follows:

1. Self -fertilized population consists of a mixture of several homozygous genotypes. Variation in such a population has a genetic component, and therefore selection is effective.
2. Any variation that occurs within a pure line is not heritable because it is due to environmental factors only. Consequently as Johannsen's bean study showed further selection within the line is not effective.

In pure line selection, selection is mainly based on the progeny performance of plants unlike phenotypic superiority in mass selection. By establishing the specific basis of progeny performance pure line theory made the process of progeny testing an essential ingradient of plant breeding programmes to determine the genetic worth of selected plants. This progeny selection was developed by Louis de vilmorin in 1856 and so it is also known as vilmorin isolation principle. Pureline selection has been the most commonly used method of improvement of self pollinated crops. Almost all the present day varieties of self pollinated crops are purelines. Pureline selection has several applications in improvement of self pollinated crops. It is used to improve Local varieties, Old pureline varieties and Introduced varieties. The pureline selection in breeding entails repeated cycles of selfing following the initial selection from a mixture of homozygous lines. The general procedure for pureline selection in developing new varieties is outlined below.

Year	Pureline selection	Steps
1st	Base Population ↓	From source population select the plants with similar and desirable characters and allowed for progeny row evaluation in the next year/season.
2nd		Select the plants from superior progeny rows and allowed to yield trials
2nd	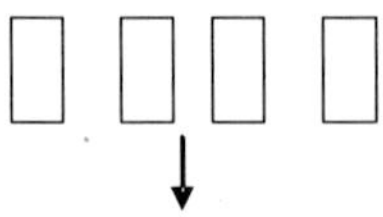	Preliminary yield trials with selected plants along with standard check. And phenotypic characters of the new variety were critically evaluated.
3rd to 5th	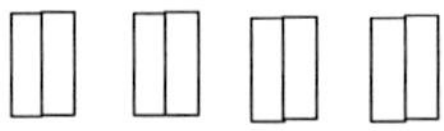	Advanced yield trials with selected plants
6th	Release best variety	Seed multiplication and distribution

Pure line method is easy and cheep method for development of extremely uniform, more attractive varieties to the farmers. Due to this extreme uniformity, the variety is easily identified in seed certification programmes. But it produces cultivars with a narrow genetic base and hence less likely to produce stable yields over a wide range of environments. Such cultivars are more prone to being wiped out by new diseases. The purity of the cultivar may be altered through admixture, natural crossing with other cultivars and spontaneous mutations. Such off type plants should be rouged out to maintain cultivar purity.

Pedigree selection: It is a method of selection, in which individual plants are selected from F_2 and the subsequent segregating generations and their progenies are tested. A key difference between pedigree selection and mass or pure line selection is that hybridization is used to generate variability unlike the other methods in which production of genetic variation is not a feature. This method was first described by H.H. Lowe in 1927. Pedigree selection is a breeding method in which the breeder keeps records of the ancestry of the cultivar. Documentation of the pedigree enables breeders to trace parent-progeny back to an individual F_2 plant from any subsequent generation.

Pedigree method is mainly applied to select the transgressive segregants suitable for improving specific characters such as disease resistance, plant height, maturity, yield and quality characters etc. It is a method of continuous individual selection after hybridization.

Generally in this method, selection starts in F_2, the generation with maximum genetic variation. In F_2 generation, fifty percent plants exhibits heterozygosity and hence selection intensity should be moderate (10%) in order to select plants that would likely include those with the desired gene combinations. The F_2 is usually space planted to allow individual plants to be evaluated for selection. The homozygosity in the F_3 is 50% less than in the F_2 and hence plants are grown in plant to row fashion for selection. As progenies become more homogeneous selection should focus more on progeny row (within and between) rather than individual plant selection in the later generations. Once selected, plants are reselected in each subsequent generation. This process is continued until a desirable level of homozygosity is attained. The general procedure for pedigree selection in developing new varieties is outlined below.

Year	Generation	Mass selection	Steps
1st	Parents	$P_1 \times P_2$	Selected parents are crossed to produce a cross product
2nd	F_1		F_1 product is space planted and harvest in bulk
3rd	F_2	Segregating Population	Plants are space planted, visual selection of superior plants and allowed to progeny row evaluation in the next year/season
4th and 5th	F_3 and F_4		Individual plant progenies are row planted, select superior plants and harvest individual plants (selection within rows practiced for two generations)
6th and 7th	F_5 and F_6		Individual plant progenies are planted in multi rows, identify superior rows, select individual plants to advance the generation (selection between rows practiced for two generations)
8th	F_7		Preliminary yield trials with selected plants along with standard check. And phenotypic characters of the new variety were critically evaluated.
9th to 11th	F_8 to F_{10}		Advanced yield trials with selected plants with more replications, over locations and years
		Release best variety	Seed multiplication and distribution

Pedigree method gives the maximum opportunity for the breeder to use his skills and judgement for selection of plants particularly in early segregating generations. Selection is based not only on phenotype but also on genotype (progeny row) making it an effective method for selecting superior lines. But selection among and within a large number of progenies in every generation is laborious and time consuming. And also maintenance of accurate pedigree records takes up valuable time. Sometimes it may be limiting factor in a large breeding programme.

Bulk method of selection: It is a method of breeding in which F_2 and subsequent generations are harvested in mass or bulk to raise the next generation. At the end of bulking period, individual plants are selected and evaluated in a similar manner as in the pedigree method of breeding. Bulk method is a strategy of crop improvement in which the natural selection effect is solicited more directly in the early generations of the procedure by delaying stringent artificial selection until later generations. This method was developed by H. Nilsson-Ehle in 1908 and later H.V. Harlan and colleagues in 1940 elaborated this method through their work in barley breeding. This method is also known as mass method or population method of breeding.

As like pedigree method this method is also applied to isolate transgressive segregants. In this method, selection of individual plants practiced in F_6/F_7 generation, by that time population approaches homozygosity i.e. waiting for opportunity for selection. The rationale for delaying artificial selection is to allow natural selection pressure to eliminate or reduce the productivity of less fit genotypes in the population. So it leads to changing of the genetic composition of population. That's why Suneson called bulk breeding as the "Evolutionary method of breeding". The general procedure for bulk method in developing new varieties is outlined below.

Year	Generation	Mass selection	Steps
1st	Parents	$P_1 \times P_2$	Selected parents are crossed to produce a cross product
2nd	F_1		F_1 product is space planted and harvest in bulk
3rd to 6th	F_2 to F_5	Segregating Population	Plants are planted at commercial seed rates and harvest plants in bulk (continue same procedure for 4 -5 generations until population approaches homozygosity) but at the end of bulking period select superior plants.
7th and 8th	F_6 to F_7		Selected plants are space planted and continue the individual plant selection for two generations and then allowed to yield trials
9th	F_8		Preliminary yield trials with selected plants along with standard check. And phenotypic characters of the new variety were critically evaluated.
10th to 12th	F_9 to F_{11}		Advanced yield trials with selected plants with more replications, over locations and years
		Release best variety	Seed multiplication and distribution

Although bulk method exploits the advantage of natural selection in terms of increasing the frequency of superior types in the population, natural selection becomes important only after F_8 or F_{10} and bulking may have to be done up to F_{20} or more. Thus the time required for developing a new variety is considerably longer than pedigree method. Pedigree records are not maintained in bulk method as like pedigree method which saves labour and time. This method provides little opportunity for the breeder to use his skills in selection. And also superior genotypes may be lost to natural selection, while undesirable ones are promoted during the early generation.

In some case, breeder needs to combine both pedigree and bulk methods, for example, selection for resistance to biotic and abiotic stresses depends on the presence of specific environments in the season, such environments do not occur every year. If environment is not favour, segregating generations may be carried in bulk until such environments occur. If favour, individual plants are then selected and handled in the pedigree method. The duration of bulking depends upon the occurrence of the concerned environment. It may end in F_2 itself or may continue up to F_6 or beyond. This method is known as mass pedigree method proposed by Harlan.

In both pedigree and bulk methods, all the genotypes are not equally represented in each generation because all plants in one generation is not advanced to the next generation. It is due to artificial selection in pedigree method and both artificial and natural selection in bulk method. So improper sampling may leads to either genetic erosion or genetic drift. This can be overcome by Single seed descent method.

Single seed descent method: It is a method of breeding in which selection of single seed from each plant present in the population and planted in bulk to raise the next generation from F_2 and subsequent generations. This method was proposed by C.H. Goulden in 1940 and later applied by Grafius (1965) in oats and Brim (1966) in soybean. Single seed descent method was came out of a need to speed up the breeding programme by rapidly inbreeding a population prior to beginning individual plant selection and evaluation, while reducing a loss of genotypes during the segregating generations.

This method is more efficient than bulk method in terms of retention of large size of variability. Because of one seed from each plant of F_2 to advance the generation under green house conditions to reach the F_5/F_6 generation that amounts to representation of F_2 spectrum of genetic variation in homozygous form. So by F_5/F_6 generation, it leads to the production of large number of homozygous progenies without sorting to selection within progenies. This method exploits the advantage of greenhouse and off season nursery facilities to advance the generations rapidly. However like bulk method, this method also suffers from lack of the provision of critical selection in early segregating generations.

Back cross method of breeding: It is a method of breeding in which F_1 and subsequent generations are back crossed with any one of the parent. This method was first proposed by H.V. Harlan and M.N. Pope in 1922. Back cross method is mainly used to replace a specific undesirable gene with a desirable alternative, while preserving all other qualities of an adapted cultivar. The adapted and highly desirable parent is called the recipient parent, while the source of the desirable gene missing in the adapted parent is called the donor parent. Recipient parent is also known as recurrent parent because it is the parent used repeatedly in the backcross programme to recover its characters. And donor parent is known as non-recurrent parent it is used only once in the breeding programme. Repeated back crossing leads to increasing the homozygosity at the same rate as like selfing. Each generation of back cross increases the homozygosity by 50% and reduces the heterozygosity by 50%. Proportion of homozygosity and homozygous individuals will be same in the population subjected to either back crossing or selfing. The proportion of homozygous individuals after 'm' number of generations of backcrossing for 'n' alleles is $(2^m - 1/2^m)^n$

Back cross is mainly applied to improve the established cultivars that are later found to be deficient in one or two specific traits by intra/inter specific transfer of simply inherited or quantitative characters. It is used to create cytoplasmic male sterile genotypes by transfer of cytoplasm one species to another. It is also used to develop isogenic lines (genotypes that differ only in alleles at specific locus) and it enables the development of multilines. Back cross programme may be modified such that F_1 may be back crossed only one or at most two times to the recurrent parent leaving much heterozygosity for transgressive segregants to appear. In back cross breeding method, the breeder crosses the recurrent parent with the donor parent. The F_1 is grown and crossed with the recurrent parent. The second step is repeated for as long as it takes to recover the characteristics of the recurrent parent. In case desired trait is dominant, a selection pressure is imposed after each back cross to identify and discard the homozygous recessive individuals. Whereas the desired trait is recessive, it will be necessary to conduct a progeny test to determine the genotype of a backcross progeny before continuing the with the next cross. The general procedure of back cross method for transfer dominant and recessive gene transfer in developing new varieties is outlined below. E.g. For convenience transfer of rust resistance gene from donor or non recurrent parent to recurrent or recipient parent with genotype was used.

Procedure of dominant gene transfer: In this recurrent parent genome is 'rr' and donor parent genome is 'RR'

Year	Generation	Crossing pattern	Remarks
1st	Parents	RR × rr ↓	Crossing between recurrent and nonrecurrent parents
2nd	F_1	Rr × rr ↓	F_1 is backcross to recurrent parent
3rd	BC_1	rr, Rr × rr ↓	Discard the susceptible plants (rr) and backcross resistant plants to recurrent parent
4th	BC_2	rr, Rr × rr ↓	Repeat same step as like above
5th	BC_3	rr, Rr × rr ↓	Repeat same step as like above
6th	BC_4	rr, Rr × rr ↓	Repeat same step as like above
7th	BC_5	rr, Rr × rr ↓	Repeat same step as like above
8th	BC_6	rr, Rr ↓	Discard the susceptible plants and self pollinate the resistant plants. Seed of selected plants harvest seperately
9th and 10th	BC_6F_2	| | | | |	Individual plant progenies grown and select the plants with rust resistance and plant type with recurrent parent (Continue the same step for two generations)
11th to 13th		↓ Preliminary and multilocation yield tirals ↓	
14th		Release new variety (Seed multiplication and distribution)	

Procedure of recessive gene transfer: In this recurrent parent genome is 'RR' and donor parent genome is 'rr'

Year	Generation	Crossing pattern	Remarks
1st	Parents	rr × RR ↓	Crossing between recurrent and nonrecurrent parents
2nd	F_1	Rr × RR ↓	F_1 is backcross to recurrent parent
3rd	BC_1	RR, Rr ↓	Self pollinate the plants for expression of recessive character
4th	BC_1F_2	RR, Rr, rr × RR ↓	Discard the susceptible plants (RR, Rr) and backcross resistant plants to recurrent parent
5th	BC_2	Rr × RR ↓	Progeny is backcross to recurrent parent
6th	BC_3	RR, Rr ↓	Self pollinate the plants for expression of recessive character
7th	BC_3F_2	RR, Rr, rr × RR ↓	Discard the susceptible plants (RR, Rr) and backcross resistant plants to recurrent parent
8th	BC_4	Rr × RR ↓	Progeny is backcross to recurrent parent
9th	BC_4	RR, Rr ↓	Self pollinate the plants for expression of recessive character
10th	BC_5F_2	RR, Rr, rr ↓	Discard the susceptible plants (RR, Rr), self the resistant plants and harvest the seeds separately.
11th and 12th	BC_5F_3	| | | | |	Individual plant progeny rows grown for two generations
13th and 14th		↓ Preliminary and multilocation yield tirals ↓	
15th		Release new variety (Seed multiplication and distribution)	

The main difference between the transfer of dominant and recessive genes is that in the later case, phenotypic identification is not possible after a cross. So each cross needs to be followed by selfing so that the progeny with the homozygous genotype can be identified and backcrossed to the recurrent parent. The genotype of new variety developed by back cross method is nearly identical with that of the recurrent parent, except for the genes transfer and also no need of extensive yield trials because the performance of recurrent parent is already known. It may save considerable time and expenses. Back cross programme is not dependent upon environment, except for the need of selection of character being transfer. Therefore, off-season nurseries and green house facilities can be used to grow 2-3 generations per year. It also saves time required for development of variety. But by the time back cross method improves the recurrent parent, it may have been replaced by other varieties superior in yielding ability and other characters.

Multiline breeding: Multilines refers to mixture of isogenic or near isogenic lines. N.F. Jensen is credited with first using this method in oats in 1952. Later N.E. Borlaug outlined systematic procedure to obtain multilines in wheat. The main objective of multiline development is disease protection to a broad spectrum of races of a disease-producing pathogen i.e. development of horizontal resistance against disease. The components of multilines, isogenic lines should be uniform for agronomic and other features except for the concerned disease resistance.

A multiline variety is usually created by mixing the seeds of several lines that are similar in appearance but have different genes for resistance to a given disease. It involves the steps like development of component (isogenic lines) lines by backcross breeding programme, evaluation of grouping of component lines in multilocation trials for yielding ability, morphologic features, agronomic characteristics and produce quality and finally releasing of multilines. Whenever disease attacks the variety, one or a few lines of the mixture would become susceptible to the pathogen in any one season. Therefore, the loss to the cultivator would be relatively low. And also the disease would spread more slowly than when the entire populations were susceptible, and reduce the damage to the susceptible line as well. Although multilines are advantageous to get horizontal resistance, the development and maintenance of multilines is a time-taking job as their component lines have to be maintained separately.

❑❑❑

Chapter-3

Breeding Methods in Cross Pollinated Crops

Cross pollinated crops characterized by high degree of heterozygosity and heterogeneity, because where normal seed set is through high degree of cross pollination. This pollination is mainly promoted by mechanisms like male sterility, self incompatability, monoecy, dioecy, heterostyly, herkogamy and dichogamy etc, which enforces each individual of a population with different genotype, heterogeneity. This encourages the breeder to exploit the heterozygous nature of individual plant either through population improvement or hybrids and synthetics development. As stated in the previous chapter, breeding methods in self pollinated crops mainly focused on individual plant improvement through progeny performance evaluation. But in cross pollinated crops, breeders focus is on population improvement i.e. improving the frequency of genes in the population for desired breeding objective.

The procedure of population improvement may be classified in several ways such as methods with/without progeny test, or unit of selection either individual plants or family of plants. Also methods may be grouped according to the population undergo selection as either intrapopulation or interpopulation. In intrapopulation improvement, selection is practiced within a specific population for its improvement for specific purposes whereas in interpopulation improvement, selection on the basis of the improvement of a cross between two populations.

S. No	Population improvement methods in cross pollinated crops	
	Criteria for classification	**Example methods**
1.	Based on progeny test	
	Without progeny test	Mass selection
	With progeny test	Ear to row method and Recurrent selection methods

2.	Based on unit of selection	
	Individual plant selection	Mass selection
	Family selection like half-sibs, full-sibs and inbred progenies etc.	Ear to row and recurrent selections
3.	Intrapopulation/ interpopulation improvement	
	Intrapopulation	Mass selection, ear to row method and recurrent selections
	Interpopulation	Reciprocal recurrent selection methods

Mass selection: Mass selection is the oldest method of selection used for population improvement in cross pollinated crops. In this, selection is mainly based on phenotypic superiority of plants. This method mainly suffers from the drawbacks like selection is based on phenotype not on real genotype so there is no control over the environment and also selection is female oriented selection because there is no control over pollination. Earlier problem can be overcome by Ear to row method and later by recurrent selection procedure. In these methods, unit of selection is family structure unlike individual plant in mass selection. In mass selection procedure, population improvement aims at improving the general population performance by selecting and bulking the superior genotypes that already exist in the population. Family selection methods mainly characterized by creation of family structure, evaluation of families and selection of superior plants by progeny testing, finally recombination of selected families to create a new base population for the next cycle of selection. These methods broadly classified in to half sib family selection methods which includes ear to row and modified ear to row methods and full sib family selection methods includes recurrent selection schemes. Half sib families mainly created by random pollination of selected female parent i.e. one parent is common for progeny families whereas full sib families created through crossing of two selected parents.

Ear to row method: It is a method of population improvement, in which seed from selected ears which have been pollinated by random pollen from the population is grown in progeny rows for the purpose of selection. This procedure was first used by C.G. Hopkins in 1899 to alter the chemical composition of corn. This is the simplest scheme of half sib family selection, in which seed from superior plants is planted in progeny row to determine the breeding value or yielding ability for any character of each plant. The seed from most productive rows or remnant seed from outstanding mother ears is bulked to complete one cycle of selection. In both the situations creation of families and recombination are carried out in the same generation. The general procedure for ear to row selection in population improvement is outlined below.

Year	Cycle	Ear to row selection	Steps
1st	One Selection cycle	Base population ↓	From source population select the plants with desirable characters, harvest plants individually and maintain the ramnant seed of each plant.
2nd		| | | | | | |	Grow progeny rows or half sibs of superior plants and select promising progenies.
2nd		Bulked population for next cycle of selection ↓	Bulk seed from superior progenies in 2 nd year or ramnant seed of selected superior plants from 1 st year grown in isolation for random mating. And the seed of this population can be grown to repeat the second cycle
May be repeated for next selection cycle			

Ear to row method is relatively simple and rapid to conduct. The main advantage of this method is progeny testing; it increases the success of selection because it is far more accurate reflection of genotype than phenotype. So environment problem can be overcome effectively. This method mainly suffers from the defect that plants in the superior progenies are pollinated by weak pollen grains; there is no control over pollination. It reduces the efficiency of selection. Even though with the use of progeny testing in this method, there is no much improvement in the population over mass selection where selection is based on individual plants. This can be achieved through management of variation due to environment and genotype-environment interaction. Lonnquist in 1964 proposed the modified ear to row method, in which open pollinated plants are planted in replicated trials over many environments in place of single row. In this evaluation of progenies is based on replicated trials. So environmental effects can be separated and the actual value of each progeny can be estimated more accurately. There are other modifications of ear to row method of selection followed by the breeder, which includes S_1 family selection, S_2 family selection and recurrent selections. In S_1 family selection, seeds for progeny testing are obtained by selfing the selected plants but not from open pollination and this is the basis for simple recurrent selection. In S_2 family selection, seeds for progeny testing are obtained after two generations of selfing. In some other cases seeds for progeny testing are obtained by crossing the selected plants to a common tester, that may be an open-pollinated variety, a hybrid or inbred and this is the basis for recurrent selection for general and specific combining ability and reciprocal recurrent selections.

Recurrent selection: Recurrent selection is a method of population improvement in which reselection generation after generation with interbreeding of selects to provide for genetic recombination. It is a cyclic selection used to improve the frequency of desirable alleles for a character in a breeding population. The idea of recurrent selection was first proposed independently by Hays and Garber in 1919, East and Jones in 1920. The procedure of recurrent selection was given by Jenkins in 1940 and term recurrent selection was coined by Hull in 1945.

This recurrent selection scheme mainly consists of three steps, which includes selfing or test crossing of individuals, evaluation of test cross or selfed progeny in replicated yield trials and finally recombination of selfed remnant seed of selected plants. This scheme is mainly devised to exploit the heterosis by isolation of desirable inbreds from the population and further utilization in production of hybrids and synthetic varieties. Based on the manner of obtaining progeny for evaluation and also specific purpose, this recurrent selection is of four different types; simple recurrent selection, recurrent selection for general combining ability (RSGCA), recurrent selection for specific combining ability (RSSCA) and reciprocal recurrent selection (RRS).

Simple recurrent selection: Simple recurrent selection is a simple intrapopulation improvement method and mainly consists of selection and selfing of superior plants, rising of progeny rows of superior plants, intercrossing of progenies in all possible combinations and finally equal amount of seed from each cross is composited to produce the next generation. This method is also known as phenotypic recurrent selection.

Year	Cycle	Simple recurrent selection	Steps
1st	One Selection cycle	Base population ↓	From source population select and self the superior plants, harvests plants individually and maintain the remnant seed of each plant.
2nd		| | | | | | |	Grow progeny rows of superior plants and made the all possible intercrosses between the progenies and finally seed from each cross is composited for next generation.
May be repeated for next selection cycle			

This method does not include any tester and it is mainly useful to improve the *perse* performance of the population. But other methods of recurrent selection includes tester, so those are suitable not only to improve performance of population but also for combining ability with a specific reference population.

Recurrent selection for general combining ability (RSGCA): It is a method of population improvement with slight deviation from simple recurrent selection that it includes tester with broad genetic base i.e. open pollinated variety or synthetic variety or segregating generation of double or multiple cross. This broad base tester is mainly used to improve the general combining ability of the population. This method is the extension of early testing of inbreds proposed by Jenkins in 1935. Early testing is the testing if inbreds for combining ability in early stages of inbreeding i.e. first or second selfed generation. These short term improved inbreds (early tested inbreds) are directly useful for synthetic varietal development.

The procedural steps of this method includes selfing and simultaneous crossing of selected plants with tester with broad genetic base, evaluations of test cross progeny in replicated trials and based on test cross progeny performance, random mating among the superior plants obtained from selfed seed. This method is effective in increasing the combining ability with open pollinated tester but not the *perse* performance of population i.e. development of reservoirs of germplasm highly suitable for the extraction of superior inbred lines.

Year	Cycle	Selection features	Remarks
1st year	One Selection Cycle	Source population × Broad base tester	From source population select superior plants, made selfing and simultaneous crossing of the plants with tester. Test cross progeny and selfed seed from each selected plant harvested separately.
2nd year			Replicated yield trials conducted for test cross progeny obtained from cross between selected plants of population and tester. And identify the superior progenies.
3rd year		Intercross block	Selfed remnant seed from plants producing superior testcross progenies planted to make all possible intercrosses. And seed from all intercrosses of a population mixed to constitute the next generation.
May be repeated for next selection cycle			

One more deviation of simple recurrent selection is improvement of specific combining ability of the population by using narrow based tester rather than broad based tester. This is possible by recurrent selection for specific combining ability (RSSCA) proposed by Hull in 1945. In this method tester is outstanding inbred line, so it permits the isolation of superior inbreds suitable for exploitation of heterosis. The procedural steps are same as like above with exception of tester.

Reciprocal recurrent selection (RRS): It is a inter population improvement method, used to improve two different heterozygous populations in the ability to combine well with each other. This method was proposed by Comstock, Robinson and Harvey in 1949. In this, each of two populations serve as testers for the plants selected from the other population. This selection method is useful in selecting simultaneously for both GCA and SCA. Selection of GCA is accomplished because the two testers (populations A and B) have broad genetic base since they are genetically heterogeneous, whereas selection for SCA is done as the two populations would be crossed with each other to produce the commercial variety, and the plants in each of the two source populations are selected for their ability to combine well with the gene combinations present in the other population, that is, for SCA with each other. The two populations developed by reciprocal recurrent selection can be used for either synthetic varietal development or for isolation of inbreds. The procedural steps of this method includes selection of superior plants from two populations, selected plants from one population are selfed and crossed as tester with selected plants in other population, evaluation of test cross progenies for both populations and remnant seed from the plants with superior testcross progenies are planted and intercrossed to reconstitute the two populations for next generation cycle.

Year	Cycle	Selection features		Remarks
1st year	One selection cycle	Population A ×	Population B	Select the superior plants from both population A and B, selected plants from one population are selfed and test crossed with several random plants from B and vice versa Test cross progeny and selfed seed from each selected plant harvested separately.
2nd year		Test cross progeny of A	Test cross progeny of B	Separated yield trials conducted for test cross progenies from populations A and B and identify the superior progenies.
3rd year		Inter cross block for Population A	Inter cross block for Population B	Selfed seed from plants producing superior testcross progenies planted separately for population A and B, make all possible intercrosses and seeds from all intercrosses of a population mixed to constitute next generation.
May be repeated for next selection cycle				

Hybrid breeding: With the predominance occurrence of heterozygosity and heterogeneity in cross pollinated crops, always encourages the exploitation of heterosis/hybrid vigour. With the same reason cross pollinated crops suffering from inbreeding depression i.e. reduction in vigour and fitness due to inbreeding. Earlier discussed population improvement schemes generally aim at keeping inbreeding low level to avoid its ill effects, but effort to exploit heterosis is rarely made. Heterosis is nothing but superiority of hybrid over parents. The term heterosis was first used by Shull in 1914. It is the basis for breeding of hybrid and synthetic varieties. In fact, hybrid varieties are best means for utilizing the heterosis while synthetic varieties utilize heterosis only partially. Hybrid variety means variety produced from cross fertilization of inbred lines with favourable combining ability. The progeny is highly heterozygous and homogeneous (F_1). An inbred line is a nearly homozygous line obtained by continuous inbreeding in cross pollinated crop. As early as 1878, Beal had shown that certain varietal crosses showed substantial heterosis and he suggested the use of such varietal hybrids. In 1909, Shull suggested the use of inbreds for development of hybrids and in 1912, East and Hayes advocated the hybrid breeding as an alternative strategy.

Development of these hybrids mainly involves the steps like production of inbreds; where inbreds produced from open pollinated plants by continuous inbreeding for 6-7 generations with accompanying selection or by culturing haploid (anther) cells followed by colchicine treatment results in perfect homozygous plants. Second step is evaluation of inbreds, which involves initially phenotypic evaluation in replicated trials leads to the discarding of inbreds with poor performance followed by evaluation for general combining ability (GCA), it means average performance of inbred in a series cross combinations with other inbreds. Top cross / poly cross test provides the reliable estimation of GCA, so that discard the inbreds with less GCA. Next level of evaluation is for specific combining ability (SCA), it means deviation of specific cross from that predicted on the basis of general combining ability of parents. Single cross test provides the reliable estimation of SCA, so select the inbreds in which hybrid combination with high SCA values. Finally best selected inbreds are used for commercial hybrid seed production. Hybrid varieties were first commercially exploited in maize. The utilization of cytoplasmic genetic male sterility is the one of significant landmarks in development of hybrid varieties.

Step 1: Production of inbred lines	Base population ↓	From source population develop inbreds by continuous selfing for 6-7 generations
Step 2: Evaluation of inbreds	Replicated trials ↓	Phenotypic evaluation in replicated trials for initial screening
	Top cross progeny test ↓	Top cross/ poly cross progeny evaluation for GCA testing
	Single cross progeny test ↓	Single cross progeny evaluation for SCA testing
Step 3: Hybrid seed production	Hybrid development	Best selected inbred lines used in crossing to produce hybrid varieties.

If two inbreds used for crossing, it is known as single cross. These single cross hybrids are more uniform but suffering from defects like lack of availability of inbreds, female inbred with low amount of seed production and male inbred with poor pollen production ability. With these problems single cross hybrid seed production is more expensive. The above mentioned defects can be overcome by double cross hybrids proposed by Jones in 1918. Double cross hybrids produced by crossing of two single crosses involving total of four inbreds, where as triple cross hybrids produced by crossing single cross with inbred involving total of three inbreds. If more than four inbred lines involved, it is known as multiple cross/complex cross used to combine monogenic characters from sources in to single genotype.

Synthetic varieties: Synthetic variety is a variety produced by crossing *interse* a number of genotypes for good general combining ability in all possible hybrid combinations, with subsequent maintenance of variety by open pollination. Hays and Garber in 1919 were apparently the first to suggest the possibility of commercial utilization of synthetic varieties. Synthetic varieties use an appreciable amount of hybrid vogour in addition to open pollination because lines used for development of synthetics compulsory tested for general combining ability. Development of synthetics involves the development of inbreds, evaluation of inbreds for general combining ability, intermating of good general combining inbreds and finally mixing F_1 seed to constitute synthetic variety. Although synthetic varieties offer a feasible means of utilizing heterosis in crop species but suffer from lack of pollination control and non viability of hybrid variety production. And the performance of synthetic varieties is always low compared to hybrid varieties because synthetics exploit only GCA, while hybrid exploits both GCA and SCA.

Composite varieties: Composite variety is a variety produced by open pollination among a number of outstanding lines usually not tested for combining ability with each other. Composites are good reservoirs of genetic variability.

❑❑❑

Chapter-4

Breeding for Biotic and Abiotic Stress Resistance

Biotic stress resistance: Breeding for biotic stress resistance/ tolerance in plants is long term objective for scientists. Plants must defend themselves against attacks from bacteria, fungi, viruses, MLOs, PLOs, invertibrates and even other plants. Because more than 42% of the potential yield is lost owing to biotic stresses, a reduction in this incidence will be one of the important targets for breeders to increase the yields while reducing the use of chemicals. Cook (2000) divided these possibilities as follows: improvement of plant material (breeding for resistance/tolerance), improvement of root health (E.g. field rotation, soil tillage, and soil borne diseases control), improvement of irrigation practices and protection against air borne diseases etc. Improvement of inherent potential of plants in the form of resistant varieties is a cheap, viable and environmental friendly alternative to reduce losses from biotic stresses. Moreover, the development of resistant varieties is an essential and predominant component of integrated pest/disease management to manage diseases and pests with minimum use of poisonous chemicals.

Plant defense mechanisms: An overview of forms of plant resistance, defined on the basis of innate and acquired resistance, and their mechanisms of response to pathogens is explained as follows. Innate resistance is exhibited by the plant in two forms: non-specific (nonhost or general) resistance and specific (host) resistance. Resistance shown by an entire plant species to a specific parasite or pathogen is known as nonhost resistance, and is expressed by every plant towards the majority of potentially pathogenic microbes. Nonhost resistance, therefore, is the most common form of disease resistance exhibited by plants. Nonhost resistance to fungi, at least, generally seems be under complex genetic control and can involve a multiplicity of defense factors that, individually, may segregate within the species without compromising overall resistance. Host resistance is usually parasite-specific in that it is restricted to a particular pathogen species, and commonly is expressed against specific pathogen genotypes. Variation in host resistance is often controlled by the segregation of single resistance (R) genes, the products of which directly or indirectly interact with 'specific elicitors' produced by the pathogen and coded for by avirulence (avr) genes. The molecular basis of nonhost resistance is poorly understood, but presumably relies on both constitutive barriers and inducible responses that involve a large array of proteins and other organic molecules produced, respectively prior to infection or during the pathogen attack. The constitutive defenses are morphological, structural and chemical barriers. An example of morphological barrier is the height of lips of stomatal guard cells. Moreover, waxes, cutin, suberin, lignin, cellulose, callose and cell wall proteins act as structural barriers that are rapidly reinforced upon the pathogen infection process. Plants also constitutively produce variety of secondary metabolites (E.g. phenolics, saponins, terpenoids, steroids and glucosignalates etc) and antifungal proteins, many of which act as antimicrobial compounds during defense. Plants employ two modes of their innate immune system to contrast pathogen infection. The first mode of immunity is referred to as pattern triggered immunity (PTI) that is triggered by molecular patterns common to many microbial types. The second mode is triggered by recognition of pathogen effectors and is called effector triggered immunity (ETI). Upon infection with necrotizing pathogens many plants develop an enhanced resistance to further pathogen attack also in the uninoculated organs. This type of enhanced resistance is referred to as systemic acquired resistance (SAR). In the SAR state, plants are primed (sensitized) to more quickly and more effectively activate defense responses the second time they encounter pathogen attack. Since SAR depends on the ability to access past experience, acquired disease resistance is a paradigm for the existence of a form of "plant memory". There is evidence that induced or acquired resistance includes the hypersensitive response (HR), a form of programmed plant cell death, cell wall strengthening, and the expression of various defense related R genes, that mediated recognition of pathogen effectors. The R genes activate a series of defense signaling cascades and pathogenesis related (PR) gene expression to generate systemic acquired resistance, this primes the plant for resistance against a broad spectrum of pathogens.

Genetic basis of resistance: The inheritance pattern of disease resistance was reported long back in 1905 by Biffen that resistance in wheat to stripe rust (*Puccinia striiformis*) was inherited as single recessive mendalian trait. Since this initial work, many genes conferring resistance to pathogens in crop plants have been characterized, and the genetic basis of pathogenicity (virulence/ avirulence) has been studied in many plant pathogens. Vanderplank (1963, 1968) classified resistance into two major types, vertical and horizontal resistance.

Vertical resistance: Specific resistance of host to the particular race of pathogen is called vertical resistance. It is also called as monogenic or oligogenic or qualitative or non uniform or major gene/ race specific or seedling or differential or complete or hypersensitive or perpendicular resistance. This resistance mainly affects the disease onset by intercepting amount of initial inoculums. Here rate of reproduction of pathogen, r = 0 for avirulent (immune reaction) type and for virulent type r = 1 (susceptible reaction). This resistance has high heritability, less durability and is easily breakdown by new race of pathogen. Influence of major genes is all the more pronounced on vertical resistance since the segregation of resistant and susceptible host is discontinuous, showing distinct phenotypic classes. Thus, simple inheritance 3:1 ratio is attendant upon vertical resistance. Vertical resistance is the true consequence of Flor's gene for gene hypothesis.

Gene for gene hypothesis: This hypothesis was proposed by Flor in linseeed rust (Melanspora lini). According to gene for gene hypothesis, for every resistance gene in the host, there must be corresponding virulence in the pathogen'. Susceptible reaction would result only when the pathogen is able to match all the resistance genes present in the host with appropriate virulence gene. In most of the pathogens, virulence is recessive to avirulence

		Host	
Pathogen		R- (resistant)	rr (susceptible)
	V- (avirulent)	R (resistant)	S (susceptible)
	vv (virulent)	S (susceptible)	S (susceptible)

The 'R' gene is susceptible to a pathogen with the corresponding virulence gene 'v' and not to the pathogen without this particular gene irrespective of how many virulence genes it may have. Thus gene R, for resistance can thus be identified only by the matching gene (v) for virulence in the form of a susceptible reaction and not by a virulent gene which does not make any distinction between resistance and susceptible genes. The main objection to this hypothesis includes, it is purely a hypothesis of gene identities, in that; it identifies the resistant gene in the host and matching gene for the pathogenesity in the pathogen. But it fails to define the quality of genes with respect to their degree of resistance and pathogenesity. Vander plank, proposed a 'second gene for gene hypothesis' which defines the genes identity and its quality both. By quality of gene, means the amount of protection that a resistance gene (identified by its locus) can confer upon its carrier against a disease. The quality of resistance gene in the host determines the fitness of the matching virulence gene in the parasite to survive when this virulence is unnecessary. When it is necessary, determines the quality of matching resistance gene, as judged by the protection it can give to the host.

Horizontal resistance: The resistance of a host to all the races of pathogen is called horizontal resistance. It is also called as minor gene or poly gene or non-specific or general or adult or field or partial or incomplete or lateral resistance. This resistance mainly affects the disease onset by reducing effect after establishment of pathogen in the host. Here rate of reproduction of pathogen; r is never zero ($r > 0$ but < 1). This resistance has low heritability and it cannot be easily overcome by new races of pathogen. This resistance is indicative of the fact that it is controlled by many genes with minor phenotypic consequences, showing a continuous pattern of segregation in segregating generations. Since segregation of resistance and susceptible genotypes is not defined by class, and that it is continuous in nature, no distinct ratios can be standardized. Suitable metric scales, usually standardized disease score indicators, are employed to measure horizontal resistance. This resistance is not applicable for gene for gene hypothesis.

Vertifolia effect: Vertifolia effect refers 'to an epidemic development in a variety carrying vertical resistance genes and low level of horizontal resistance genes, heading to heavy economic losses. Vander plank introduced the term vertifolia effect derived from the name of a German potato variety having the late blight resistance genes R_3 and R_4. The variety became susceptible when the appropriate pathogen (P_3, P_4) evolved in the pathogen leading to complete failure of crop. The failure is mainly due to level of horizontal resistance in the varieties carrying oligo genes for resistance is usually low and the pathogen is able to evolve the virulent phenotype. Parlevilet (1981) reported vertifolia effect is not universal, as Barley variety 'cebada capa' carrying P_7 (resgene-V_R) also having high level of polygenic partial resistance to leaf rust. To avoid vertifolia effect, one should try to reject susceptible one in place of selecting highly resistance plant. Kumar et.al (1977) suggested combining both horizontal resistance (HR) and vertical resistance (VR) and first select for HR, crossing with VR and selecting further in segregating generation.

Boom and Bust cycle: When a new variety carrying VR (vertical resistance) is development it is resistant to the prevalent pathotype of the pathogen in question, therefore it becomes popular rapidly and occupy a large area under the concerned crop this period of boom continuous till the pathotype specific to the VR gene employed in the variety evolves and / or increase in frequency to cause on epidemic in the fields occupied by the variety. Now susceptible variety loss popularity and area under it declines sharply thus the variety is effectively busted by the virulent pathotype. This has been called as boom and bust cycle by Priestley (1970).

Mechanisms of disease resistance: There are different ways of disease resistance *viz.*, disease escape, disease endurance or tolerance disease resistance and immunity

i. **Disease escape:** The ability of susceptible host plants to avoid attack of disease due to environmental conditions factors is called disease escape. The factors responsible for escape are early varieties, charge in the date of plating, change in the site of planting; balanced application of NPK etc. E.g. Early varieties of groundnut and potato may escape 'tikka' and 'late blight' diseases respectively since they mature before the disease epidemic occurs, changing planting season in sugarcane from June to October has successfully escaped leaf-rust.

ii. **Disease endurance or tolerance:** The ability of the plants to tolerate the invasion of the pathogen without showing much damage is called disease tolerance. This endurance is brought about by the influence of external characters. Generally, tolerance is difficult to measure since it is confounded with partial resistance and disease escape. To estimate tolerance the loss in yield and some other trait of several host varieties having the same amount of disease like leaf area covered by disease etc., is compared. E.g. Wheat varieties when fertilized with potash and phosphorus are more tolerant to the rust and mildew infection.

iii. **Disease Resistance:** The ability of plants to withstand, oppose or overcome the attack of pathogens is resistance. Resistance is a relative term and it generally refers to any retardation in the development of the attacking pathogen. In case of resistance, disease symptoms to develop and the rate of reproduction 'r' is never zero but it is sufficiently lower than one (the rate of reproduction on the susceptible variety) to be useful. Resistance is largely controlled by inherited characters may be controlled by single dominant gene in Ottawa 770 B, Newland flax variety, wheat all rusts NP 809.

iv. **Immunity**: When the host does not show the symptoms of disease it is known as immune reaction. Immunity may result from prevention of the pathogen to reach the appropriate parts of the host E.g. exclusion of spores of ovary infecting fungi by closed flowering habit of wheat and barley. It is more generally produced by hypersensitive reaction of the host usually immediately after the infection was occurred. In immune reaction the rate of reproduction in zero i.e. r = 0

v. **Hypersensitivity:** Immediately after the infection several host cells surrounding the point of infection are so sensitive that they will die. This leads to the death of the pathogen because the rust mycelium cannot grow through the dead cells. This super sensitivity (hypersensitivity) behaves as a resistant response for all practical purposes. Phytoalexins are specific polyphenolic or terpenoid chemicals and are produced by the host in response to the infection by a pathogen. More than 30 different phytoalexins have been identified. Phytoalexins are either fungicidal or fungistatic.

Factors for causing disease resistance: The disease resistance may be caused due to morphological, physiological, anatomical, nutritional and environmental factors. Morphological features includes structural and functional characteristics mainly prevent the entrance of pathogen i.e. prevents the first stage of infection. For example resistance to jassid attack in cotton has been shown to be correlated with the hariness of varieties; hairy type resists the attack more, than glabrous types. Young sugarbeet leaves practically immune to attack of the circospora because the stomata size is very small. Physiological features, by virtues of its chemical composition the protoplasm may exert an inhibitory influence on the pathogen bringing about the desired resistance in the plant; E.g. Resistance of grape to powdery mildew is highly correlated with the acidity of cell sap. Presence of toxic substance in the red pigment in the coloured onions. The outer scales resist the smudge fungus attack when the scales are removed they become susceptible. Anatomical features like more secondary thickening of the cell walls of resistant potato varieties which resists the mechanical puncture of the invading *Pythium* pathogen. Nutritional factors includes reduction in growth and in spore production is generally supposed to be due to unfavourable physiological conditions within the host. Most likely a resistant host does not fulfill the nutritional requirements of the pathogen and thereby limits its growth and reproduction. Environmental factors like temperature, moisture, humidity and soil p^H and fertility status of the soil effects the pathogen reaction greatly.

Methods of breeding for disease resistance: Commonly used methods for disease resistance includes

i. Introduction: Resistant varieties may be introduced for cultivation in a new area. E.g. early varieties of groundnut introduced from USA have been resistant to leaf spot (Tikka), Kalyanasona and Sonalika wheat varieties originated from segregating material introduced from CIMMYT, Mexico, were rust resistant.

ii. Selection: Selection of resistant plants from commercial varieties is easiest method. E.g. MCU I was selection from CO4 for black arm resistance in cotton, Pusa Sawani of bhendi (resistant to yellow mosaic) is a selection from a collection obtained from Bihar.

iii. Hybridization and selection: Pedigree method is quite suitable for horizontal resistance. Artificial disease epiphytolics are produced to help in selection for disease resistance. E.g. Laxmi in Cotton (Gadag 1 x CO2) for leaf blight resistance. Another, back cross method is used to transfer resistance genes from an undersirable agronomic variety to a susceptible, widely adoptable and is agronomically highly desirable variety. If the resistant parent is a wholly unadapted variety, backcross method is a logical choice. If resistant variety also possess some good qualities then chose pedigree method of handling segregating material.

iv. Budding and grafting: The disease resistance in vegetatively propogated material is transferred by adopting either by budding or grafting. By grafting or budding the resistant material, the resistance can be transferred.

v. Mutation breeding: When adequate resistance is not available in the germplasm; Mutation breeding is resorted to induce resistance.

vi. Transgenic breeding: Genetic engineering for disease resistance has been very limited. But by using the approaches like antifungal protein mediated resistance and antifungal compound mediated resistance are commonly used to develop fungal resistant transgenic line. E.g. Transgenic potato plants expressing an H_2O_2 generating fungal gene for glucose oxidase were found to have elevated levels of H_2O_2 and enhanced levels of resistance to both fungal and bacterial pathogens particularly *verticillium* wilt.

Like disease infestation, the incidence of insect pests is another form of biotic stress that lowers economic performance of crop plants. To overcome this problem modern agriculture uses a wide range of insecticides to control the pest damage. However, chemical control of pests, in addition to being expensive, frequently results in negative environmental effects. But the development of insect resistant varieties is therefore an important plant breeding strategy. In this section attention will mainly given to mechanisms promoting insect resistance in plants.

Mechanisms of pest resistance: Insect resistance is grouped in to four categories: non preference, antibiosis, tolerance and avoidance. The first three types of resistance were recognized by Painter (1951), while the fourth, avoidance, was included subsequently.

i. **Non preference:** Host varieties exhibiting this type of resistance are unattractive or unsuitable for colonization, oviposition or both by an insect pest. This type of resistance is also termed as nonacceptance or antixenosis. Nonacceptance appears to be more appropriate, since in this type of resistance, insects will not accept a resistant host variety even if there is no alternative source of food. Non preference involves various morphological and biochemical features of host plants such as - color, hairness, leaf angle, taste etc.

ii. **Antibiosis:** It refers to an adverse effect of feeding on a resistant host plant on the development and or reproduction of the insect pest. In severe cases, it may even lead to the death of insect pest. Antibiosis may involve morphological, physiological or biochemical features of the host plant; some cases of insect resistance involve a combination of features. E.g. Resistance to brown plant hopper in rice is due to antibiosis and non preference.

iii. **Tolerance:** An insect tolerant variety is attacked by the insect pest to the same degree as a susceptible variety. But at the same level of infestation, a tolerant variety produces a larger yield than a susceptible variety. Ability of the host plant to withstand the insect population to a certain extent which might have damaged a more susceptible host. Tolerance is mainly a host character and it may be because of greater recovery from pest damage. E.g. Rice varieties tolerant to stem borer/gall midge produce additional tillers to compensate yield losses

iv. **Avoidance:** Insect avoidance result from the host plants being at a much less susceptible developmental stage when the pest population is at its peak. Pest avoidance is the same as disease escape, and as such it is not a case of true resistance. E.g. early maturing cotton varieties escape pink bollworm infestation, which occurs late in the season.

Factors for causing pest resistance: The main factor for insect resistance includes morphological, physiological and biochemical features of host plant. Morphological factors like, hairiness, colour, thickness and toughness of tissues etc. are known to confer insect resistance. Hairiness of leaves is associated with resistance to many insect pests leaf beetle in cereals, in cotton to Jassids, in turnip to turnip aphid. Colour of plant may contribute to non preference in some cases. E.g. Red cabbage, Red leaved brussel's sprouts is less favored than green varieties by butterflies and certain Lepidoptera for oviposition. Boll worms prefer green cotton plants to red ones. Thickness and Toughness of plant prevent mechanical obstruction to feeding and oviposition and thereby lead to non-preference as well as antibiosis. E.g. thick leaf lamina in cotton contributes to jassid resistance; solid stem in wheat confers resistance to wheat stem sawfly. Physiological factors like osmotic concentration of cell sap, various exudates etc may be associated with insect resistance. E.g. Exudates from secondary trichomes of *Medicago disciformis* leaves have antibiotic effects on alfalfa weevil. In cotton, high osmotic concentration of cell sap is associated with jassid resistance. It is believed that biochemical factors are more important than morphological and physiological factors in conferring non-preference and antibiosis. E.g. High concentrations of gossypol is associated with resistance in several insect pests in cotton, high silica content in shoots gives resistance to shoot borer in rice.

Methods of breeding for insect resistance: The various methods of breeding for insect resistance includes

i. Introduction: E.g. Introduction of resistance grape root-stocks for *Phylloxera vertifoliae* from U.S.A. into france.

ii. Selection: E.g. Resistance to potato leaf hopper

iii. Hybridization and selection: Pedigree and back cross methods are commonly followed.

iv. Transgenic breeding: Genetic engineering has received maximum attention for the development of resistance to insect. Several insect resistant transgenic varieties have been tested and released in crops like tomato, potato, cotton and maize. There are approaches like introduction resistant genes from micro organisms or from higher plants have been adapted to develop insect resistant transgenic plants by transferring insect control protein genes. E.g. Btcotton, where cry gene of *Bacillus thuringenesis* was transferred to cotton plant against boll worm resistance.

But breeding for insect resistance suffers from several problems like

i. Breeding for resistance to one insect leads to the susceptibility to another pest. This may be because the feature of host plant associated with resistance to one insect is associated with susceptibility to another insect.

   a. E.g. Glabrous (hairless) strains of cotton are resistant to bollworms, but they are susceptible to jassids.

ii. Breeding for insect resistance reduces the quality of produce or may even make it unfit for consumption.

   a. E.g. High gossypol content is associated with resistance to bollworm in cotton.

iii. Genes for insect resistance are available only in the related wild species. So problem with linkage drag i.e. linkage between the genes for resistance and some undesirable genes.

iv. Screening for insect resistance is the most critical and difficult step in a breeding programme for insect resistance. This necessitates a close coordination between the breeder and the entomologist.

v. Breeding for insect resistance is a long term programme. Therefore, adequate financing over a long period of time with a follow-up action is essential.

As in many years of the past, in addition to diseases and pests, plant viruses are forming a major problem in the cultivation of many agricultural crops throughout the world. For long times, these pathogens have been controlled using conventional measures like crop rotation and other cultivation techniques, early detection, destruction of infected source plants, cross-protection, breeding for resistance, and chemical control of their vectors. Increased knowledge of both the molecular genetics of plant viruses and, more recently, also of their hosts' natural defence systems have resulted in the development of a number of novel ways to control virus diseases in plants. In this section more attention given to mechanisms promoting virus resistance.

Natural resistance of plants to viruses: Besides passive defence based on the presence of existing barriers like the rigid cell wall, plants exhibit active defence mechanisms upon recognition of pathogens such as viruses. The most common mechanism associated with active defence is the so-called hypersensitive response (HR), during which cells surrounding the primary infection site of the virus die due to a rapidly induced programmed cell death, which results in a visible necrotic local lesion. The induction of this response is preceded by a specific recognition of the virus, and in many cases this is based on matching (dominant) gene products of the plant (produced from dominant resistance genes, R genes) and the virus (avirulence genes). R genes in palnts are defined by the classical gene for gene hypothesis, which states that for every incompatible host pathogen interaction, there exist matching R genes in the host and *avr* genes in the pathogen. Resistance reaction against pathogen results generally by direct interaction between the products of R and *avr* genes. All known R genes encode products having two basic functions: to act as sensors for the corresponding *avr* factors/ elicitors and to initiate signaling cascades for the expression of defense related genes. A number of structural features are conserved across several R gene products includes leucine-rich repeat (LRR), nucleotide-binding site (NBS), serine-threonine kinase, leucine zipper and toll-interleukin region (TIR) etc. E.g. N gene of tobacco provides resistance against TMV. And the N gene product has a prominent TIR signaling domain at the amino-terminus and a LRR a recognition domain at the carboxyl-terminus of the polypeptide.

Pathogen derived resistance (PDR), in this a part or a complete viral gene is introduced into the plant, which subsequently interferes with one or more essential steps in the life cycle of the virus. This can be achieved by either viral coat protein or replicase gene encoding sequences or movement proteins or *Satellite* RNAs.

Coat protein mediated resistance (CPMR): The expression of viral coat protein gene in transgenic plants was shown to induce protective effects similar to classical cross protection, and was therefore distinguished as "coat protein mediated protection". The level of protection conferred by CP genes in transgenic plants varies from immunity to delay and attenuation of symptoms. Powell-Abel *et al* in 1986 first reported resistance against TMV in transgenic tobacco expressing the TMV CP gene.

Replicase mediated resistance: Engineering virus resistance by using genes encoding viral RNA- dependent RNA-polymerases (RdRps) is called replicase mediated resistance. Replicase (Rep) protein mediated resistance against a virus in transgenic plants was first shown in tobacco against TMV in plants containing the 54 KDa putative *Rep* gene.

Movement protein mediated resistance: Movement proteins (MP) are essential for cell to cell movement of plant viruses. These proteins have been shown to modify the gating function of plasmodesmata, thereby allowing the virus particles or their nucleoprotein derivatives to spread to adjacent cells. This phenomenon was first used to engineer resistance against TMV in tobacco by producing modified MP which is partially active as a transgene.

Satellite RNA mediated resistance: The presence of sat-RNA modulates the symptoms induced by the helper virus and often depresses helper virus accumulation in different host species. CMV sat RNA depends on its helper virus CMV for replication, movement within the plant, encapsulation and transmission. Transgenic tobacco plants expressing multiple or partial copies of CMV sat-RNA showed attenuated symptoms when challenged with CMV. In addition, tobacco plants transformed with anti-sense sat-RNA showed delayed symptom development with the cognate virus.

Non pathogen derived resistance: Strategies for non-pathogenic derived resistance mainly includes utilizing genes derived from either the host plant or any other non-pathogenic source. The other non-pathogenic derived strategies are the utilization of plant disease resistance genes, the ribosome –inactivating proteins, plant proteinase inhibitors, and human interferon-like systems, antiviral antibodies expressed in plants, systemic acquired resistance and secondary metabolic engineering.

Ribosomal inactivating proteins: Ribosomal inactivating proteins (RIPs) inhibit the translocation step of translation by catalytically removing a specific adenine base from 28S ribosomal RNA. They are synthesized either as pre or pre-pro proteins and targeted to vacuoles. Because of their specific intracellular localization, RIPs donot affect the endogenous 28S RNA. RIPs enter in to the cells together with the viruses and exert the damage to the host ribosome or possibly viral RNA. When RIPs are mixed with viruses and applied on plants, virus multiplication and symptom development are dramatically suppressed. RIPs not only inhibit local virus multiplication in RIP-treated leaves but also block viral multiplication systematically. E.g. Transgenic *Nicotiana benthamiana* plants expressing PAP (Pokeweed) have been shown to offer broad-spectrum virus resistance to both mechanical and aphid transmission.

Protease inhibitors from plants: Many viruses, namely poty-, tymo-, nepo-, como-, and closteroviruses need cysteine protease activity to process their own polyproteins for their replication and propagation. E.g. Tobacco lines expressing the rice cysteine protease inhibitor gene were examined for resistance against tobacco etch virus (TEV) and PVY infection.

Interferon like systems: Higher vertebrates resist virus infections in part by catalysis of RNA decay using the interferon regulated 2-5A system. The 2-5A synthetase that makes 5′ phophorylated, 2′-5′- linked oligo adenylates (2-5A) in response to double stranded DNA, and the 2-5A dependent RNAse L. In plants, homologues of this system are not yet known but the inducers, i.e. interferon-like molecules have been reported. E.g. The transgenic tobacco produced low-level but functional 2-5A synthetase and activated RNAse L. these transgenic lines were tested positive for their proficiencies to resist at least three different types of viruses: TEV, TMV and AMV.

Anti-viral plantbodies: Expression of specific anti-viral antibodies in plants to control plant viruses are commonly known as plantbodies. A panel of monoclonal antibodies was raised against ACMV and the gene for the most reactive of the above panel was cloned and expressed in *Nicotiana benthamiana.*

Systemic acquired resistance: Plants develop active resistance first at active localized site only, but spreads systemically in due course. This resistance is called systemic acquired resistance (SAR). SAR is characteristically associated with accumulation of salicylic acid (SA), enhanced expression of pathogen related (PR) proteins leading to the synthesis of higher phenolic compounds, increase of active oxygen species and reinforcement of cell wall by the deposition of lignin and suberin. E.g. Involvement of SA in TMV resistance has been shown by expressing the bacterial salicylate hydroxylase (NahG) gene in tobacco palnt, thus decreasing its endogenous salicylic acid, and causing susceptibility to TMV infection.

Secondary metabolite pathways: Metabolic pathways are important in viral pathogenesis as they are key targets for intervention against viral infection. S-adenosyl homocysteine hydrolase (SAHH), is key enzyme in trans-methylation reaction. The antisense RNA for tobacco SAHH was expressed in transgenic tobacco plants. Though 50% of the plants showed stunting, they were resistant to infection by various plant viruses.

A new phenomenon called post-transcriptional gene silencing (PTGS) has recently been shown to be responsible for the inherent ability of many plants to specifically degrade nucleic acids in a sequence specific manner.

Post-transcriptional gene silencing (PTGS): It is a specific RNA degradation mechanism of any organism that takes care of aberrant, unwanted excess or foreign RNA intra-cellularly in a homology-dependent manner. This activity could be present constitutively to help normal development or induced in response to cellular defense against pathogens. In this mechanism, the elicitor double-stranded RNA (ds RNA) commonly produced during viral infection is degraded to 21-25 nucleotides, termed as small interfering RNA (siRNA). When viral RNA is either the elicitor or target of PTGS, the degradation mechanism is known as virus induced gene silencing (VIGS). VIGS comes into play when plants recover from initial viral infection or plants resist super infection of viruses with genomes bearing homology with those of the viruses used as primary inoculums. E.g. If tobacco rattle virus (TRV) infects *Nicotiana benthamiana,* the plant develops initial symptoms of viral infection at the inoculated region. But the plant shows signs of recovery later and newly emerged leaves are free of TRV.

Abiotic stress resistance: Now-a-days, agriculture has new and huge challenges, due to population growth, the pressure on agriculture liability on the environmental conservation, and climate change. There are several environmental factors that have adverse effects on normal growth ad development of crop plants. Such factors include deficiency or toxicity of minerals, moisture deficit and low temperature, soil salinity and alkalinity. These factors are known as abiotic or no biotic factors. The various non biological factors of environment interferes with the complete expression of genotypic potential is called abiotic stress. Abiotic stresses result due to moisture, temperature, mineral, salinity, soil p^H, air pollution, etc. In this chapter more importance given to abiotic stresses like drought, water logging, salinity, chilling and freezing etc.

Drought: Drought may be the "inadequacy of water availability, including precipitation and soil moisture storage capacity, in quantity and distribution during the life cycle of a crop to restrict the expression of its full genetic yield potential." It mainly affects 26 % of the arable area. It may be pointed out that the relative importance of different abiotic stresses varies greatly depending on the geographical region and the specific location within a given region. For example, the degree of moisture stress experienced by a crop will be greater under warmer conditions than that under relatively cooler environments, the soil moisture level and air humidity's being comparable under both the environments. In addition, the severity of the effects of a stress on a crop depends greatly on the duration of stress and on the stage of crop growth at which the stress has occurred.

Mechanism of drought resistance: There are different ways of drought resistant mechanisms *viz* drought escape, drought avoidance, drought tolerance and drought resistance.

i. **Drought escape:** It is mainly due to ability of a genotype to mature early, before occurrence of drought. Drought escape is most common in plants of desert region. E.g. Early maturing varieties of sorghum, maize, bajra, wheat, rice etc; give more yield than late maturing under drought.

ii. **Drought avoidance:** It is due to the ability of plants to maintain favourable water balance even under stress. The plants which avoid drought retain high moisture content is their tissues and lose less water. This is possible either because of increased water uptake (due to increase in root development) plants are called water spenders or reduced water loss (due to reduction in growth of aerial parts are called water savers i.e. to avoid transpiration. Several traits such as leaf rolling, folding and reflectance narrow leaves, increased pubescence on aerial organs, presence of awns, osmatic adjustment of stomata, cuticular wax mainly contributes drought avoidance. Reduced transpiration is mainly by increasing in concentration of Abscisic Acid (ABA), closure of stomata, ABA plays role in reduction of leaf expansion, Promotion of root growth etc.

iii. **Drought tolerance:** Ability of plants to produce higher yield even under 'low water potential'is called drought tolerance. Tolerant cultivar generally exhibits better germination, seedling growth and photosynthesis. Drought tolerance may be because of high proline accumulation or maintenance of membrane integrity

iv. **Drought resistance:** It is the sum total of avoidance and tolerance. It refers to the genetic ability of plants to give good yield under moisture stress conditions.

Factors for causing drought resistance: The drought resistance may be caused due to morphological, physiological and biochemical factors. Morphological features include earliness, reduced tillering, leaf characters like leaf rolling, leaf folding, leaf shedding, leaf reflectance, reduced leaf area, narrow leaf, change in leaf angle, hairiness (presence of hairs on leaf and other parts, lowers leaf temperature and reduce transpiration), colour of leaves, wax content, awns, root system (rooting depth and intensity). Physiological parameters includes photosynthesis (efficient system like C4) under stress, photosynthetic efficiency is reduced due to chloroplast damage, reduced transpiration and reduced respiration losses, stomatal behavior (closure of stomata, also change in size and number of stomata), osmotic adjustment, leaf enlargement (increase in thickness) and leaf cuticle wax etc. biochemical factors includes accumulation of proline and betaine, increase in Abscisic acid (barley) and Ethylene (maize & wheat), protein synthesis (increases under stress) and nitrate - reductase activity etc.

Methods of breeding for drought resistance: Breeding for drought resistance refers to breeding for yield under moisture stress, i.e. developing varieties which can give high yields under stress. It is important that drought resistance be incorporate in material with high genetic potential for yield. The commonly used breeding methods includes introduction, selection, hybridization followed by selection, mutation breeding and transgenic approaches etc.

Water logging: As per Levitt (1980) flooding (i.e. water logging) is the presence of water in soil excess of field capacity. Once soil becomes water logged, air space in soil is displaced with water, the O_2 in the soil in dissolved in water. i.e. O_2 decreases, Co_2, ethylene and other toxic gases increases. By this, root systems are suddenly plunged into an anaerobic condition. This switching from aerobic to anaerobic respiration disrupts root metabolism. Because of this carbohydrates level get depleted due to dissipation of metabolism, high water temperature and low light. Finally it leads to the reduced growth / elongation, chlorosis, senescence and abscission of lower leaves, wilting and leaf curling, hypertrophy (increase in size of organ due to increase in cell size) and epinasty (downward growth of petioles) of plants. So in this section attention will mainly given to mechanisms promoting resistance/tolerance to water logging.

Mechanisms promoting tolerance/resistance to water logging are

i. Adventitious root formation on lower part of stem (close to surface so that O_2 tension is quickly restored after transient water logging)

ii. Lenticel (i.e. raised pores in the stem of plants) formation

iii. Aerenchyma formation (soft plant tissue continues air spaces found in acquatic plants) in the cortex that provide canal paralled to the axis of the root through which gases can diffuse longitudinally.

iv. Elongation capacity (In rice - best elongation response give 100% recovery from submergence and poorest elongation gives upto 49% recovery) Scoring for elongation can be done between booting and flowering stage after flooding the crop to varying depths.

Breeding methods for water logging tolerance/resistance are same as like drought.

Salt stress/salinity: It is nothing but osmotic forces exerted on plants when they are growing in a salt marsh or under other excessively saline conditions. While salt tolerance mainly refers to the ability of plants to prevent, reduce or overcome injurious effects of soluble salts present in their root zone. It is a global problem as saline and alkali soils are found in almost all the countries of the world, more in Semi Arid Tropic (SAT) of world. The main symptom of plants during salt stress includes retardation/cessation of growth, necrosis, leaf abscission, loss of turgor and ultimate death of plant. Problem of salinity can be overcome by either soil reclaimation, but it is costly, time consuming and short lived or through resistant varieties, it is less costly, more effective, long lasting require longer period to develop.

A mechanism promoting salt stress tolerance includes salt tolerance and salt avoidance. Salt tolerance is mainly by response of plants to salinity stress by accumulating salt, generally in their cells or glands and roots etc. Salt avoidance is mainly by maintaining their cell salt concentration unchanged either by water absorption (e.g. Rice) or by salt exclusion (e.g. tomato, soybean, citrus, wheat grass). Breeding methods for salt tolerance are same as like drought but breeding for yield potential should have greater emphasis than breeding for salt resistance per se because screening is done on the basis of yield reduction in stress environment as compared to non-stress Environment.

Cold stress: This stress occurs at temperatures less than 50°F (10°C), varying with chill factor, wetness, protection from wind. Temperature is one of the important factors of this stress, which determine the distribution of plants geographically in an optimum environment where they can survive and complete their life cycle. When temperatures remain above-freezing i.e. $>0^0$C to $<10^0$-15^0C it is called chilling. When temperature remain below freezing i.e.$<0^0$C it is called Freezing.

Chilling tolerance: Ability of some genotypes to survive / perform better under chilling stress than other genotypes is called chilling tolerance. Chilling sensitive plants are typically tropical plants. Temperate plants are generally tolerant to chilling injury. Chilling sensitive plants mainly shows the effects like reduced germination, poor seedling establishment, stunted growth, wilting, chlorosis, necrosis, pollen sterility, poor fruit set / seed formation, reduced root growth, locked open stomato, ABA accumulation, reduced membrane stability, poor chlorophyll synthesis (affected), reduced photosynthesis & respiration and toxicity due to H_2O_2 formation. Chilling tolerance is mainly because of chilling hardening, *i.e.* an earlier exposure to a near chilling temperature for a specified period as a result of which chilling tolerance of the concerned plants increases. The main mechanisms promoting chilling tolerance are

i. Membrane lipid un-saturation

ii. Reduced sensitivity of photosynthesis

iii. Increased chlorophyll accumulation

iv. Improved germination

v. Improved fruit / seed set

vi. Pollen fertility

Freezing tolerance: It is the ability of plants to withstand subzero temperatures through the formation of ice crystals in the xylem and intercellular space, or apoplast of their cells. Freezing tolerance is enhanced as a gradual adaptation to low temperature through a process known as cold acclimation, which initiates the transition to prepare for subzero temperatures through alterations in rate of metabolism, hormone levels and sugars. The main effects of freezing are ice formation at both intra and inter cellular levels. Freezing mainly causes disruptions is and / or alter the semipermeable properties of plasma membrane, loss of solutes from the cells occur, cells remain plasmolyzed even after thawaing which is often called as frost plasmolysis and cells may become highly turiel due to uptake of excess water. Cooling of water below 0°C without ice crystal formation is called supercooling. Freezing resistance is a complex trait involving physiological, chemical & physical processes at the tissue and cell level.

The main mechanisms promoting freezing resistance are freezing avoidance and freezing tolerance. Freezing avoidance is the ability of plant tissues/or genes (but the whole plants) to avoid ice formation at sub zero temperature is called freezing avoidance. Supercooling is a mechanism of freezing avoidance it is controlled by lack of ice-nucleators, small cell size, little or no intercellular space, low moisture content, barriers against external nucleators and presence of antinucleators. Freezing tolerance is the ability of plants to survive the stresses generated by extra cellular ice formation and to recover and regrow after thawing is known as freezing tolerance. The various components of freezing tolerance are osmotic adjustment, amount of bound water, plasma membrane stability, cell wall components properties and cold-responsive proteins (e.g. ABA).

❒❒❒

Chapter-5

Breeding in Cereals: Rice, Wheat and Maize

Rice

Rice *(Oryza sativa* L.*)*, a model cereal species, is the primary food source for more than half of the world's population. More than 90% of rice is produced and consumed in South and South-East Asia and it is an important food crop for Africa, Latin America and Middle East. The genus to which it belongs is *Oryza* and the tribe Oryzeae of the family Gramineae (Poaceae). The genus *Oryza* contains 25 recognized species, of which 23 are wild species and two, *O. sativa* and *O. glaberrima* are cultivated. *O. sativa* is the most widely grown of the two cultivated species. It is grown worldwide including in Asian, North and South American, European Union, Middle Eastern and African countries. However, *O. glaberrima* is grown solely in West African countries.

The centre of origin and centres of diversity of two cultivated species *O. sativa* and *O. glaberrima* have been identified using genetic diversity, historical and archaeological evidences and geographical distribution. It is generally agreed that river valleys of Yangtze, Mekon rivers could be the primary centres of origin of *O. sativa* while Delta of Niger River in Africa as the primary centre of origin of *O. glaberrima.* The foothills of the Himalayas, Chhattisgarh, Jeypore Tract of Orissa, northeastern India, northern parts of Myanmar and Thailand, Yunnan Province of China etc., are some of the centres of diversity for Asian cultigens. The Inner delta of Niger river and some areas around Guinean coast of the Africa are considered to be centre of diversity of the African species of *O. glaberrima.*

The views regarding the progenitor of rice can be grouped in to two classes *viz.,* Polyphyletic and Monophyletic origin. According to polyphyletic origin, *O. sativa* and *O. glaberrima* are believed to have evolved independently from two different progenitors, viz. *O. nivara* and *O. barthii* and they are believed to be domesticated in South or South East Asia and tropical West Africa respectively. The progenitors of *O. sativa* are considered to be the Asian AA genome diploid species and those of *O. glaberrima* to be African AA genome diploid species *O. barthii* and *O. longistaminata.* According to monophyletic origin, both Asian rice and African rice arose from a common parent *O. perennis.* This view is the most accepted one because both Asian rice and African rice are similar except in glume pubescence, ligule size and colour of pericarp.

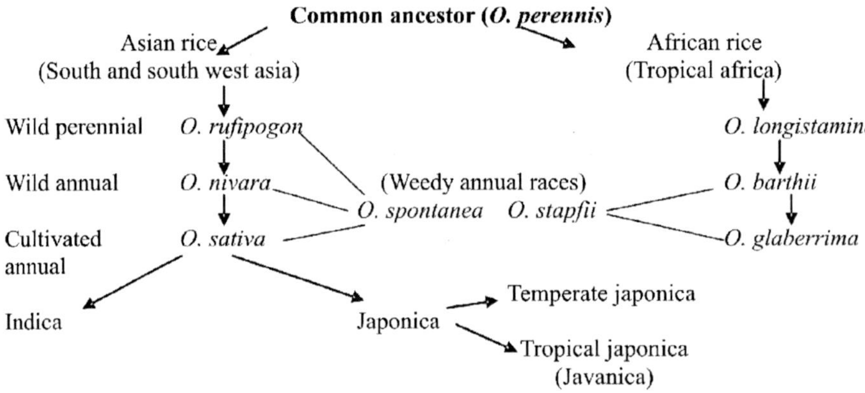

Species of the genus *Oryza*: The species of the genus *Oryza* are broadly classified into four complexes (Vaughan, 1994) viz. Sativa, Officinalis, Ridley and Meyeriana (Table). Of these, Sativa and Officinalis complexes are the best studied. The Sativa complex comprises the cultivated species *O. sativa* and *O. glaberrima* and their weedy/wild ancestors viz., perennial rhizomatous *O.longistaminata, O.barthii* (formerly *O. breviligulata*) and *O. rufipogon, O. nivara* and *O. sativa f. spontanea.* The basic chromosome number of the genus *Oryza* is 12. *O. sativa, O. glaberrima* and 14 wild species are diploids with 24 chromosomes, and eight wild species are tetraploids with 48 chromosomes. Genome analysis done on the basis of chromosome pairing behaviour and fertility in interspecific hybrids and degree of sexual compatability, has made possible to group them under nine distinct genomes, viz., A, B, C, D, E, F, G, H and J. On the basis of crossability and ease of gene transfer, the primary gene pool of rice is known to comprise the species of Sativa complex, while the species belonging to Officinalis complex constitute the secondary gene pool. Crosses between *O. sativa* and the species of Officinalis complex can be accomplished through embryo rescue technique. The species belonging to Meyeriana, Ridleyi complexes and *O. schlechteri* constitute the tertiary gene pool (Chang, 1964).

Species complex	Chrom- number	Gen- ome	Distribution	Useful traits
Sativa complex				
O. sativa	24	AA	World wide	Cultigen and high yielding
O. nivera	24	AA	Tropical and subtropical Asia	Resistant to grassy stunt virus and BLB
O. rufipogon	24	AA	Tropical and subtropical Asia, Tropical Australia	Source of CMS, and Resistant to BLB, blast, BPH, tungro virus
O. meridionalis	24	AA	Tropical Australia	Elongation ability and drought avoidance and tolerance to heat and drought
O. glumaepetula	24	AA	South and Central America	Elongation ability, source of CMS and tolerance to heat
O. glaberrima	24	AA	West Africa	Cultigen, resistant to BLB, blast, African gall midge, nematodes and weed competitiveness
O. barthii	24	AA	Africa	Resistance to BLB and GLH, drought avoidance , tolerance to heat and drought
O. longistaminata	24	AA	Africa	Resistant to BLB, nematodes, stem borer and drought avoidance
Officinalis complex/ Latifolia complex				
O. punctata	24	BB	Africa	Resistant to BPH, BLB, Zigzag leaf hopper, tolerance to heat and drought
O. rhizomatis	24	CC	Sri Lanka	Drought avoidance, resistance to blast and tolerance to heat

O. malamp-huzaensis	48	BBCC	South East Asia	
O. minuta	48	BBCC	Philippines and New guinea	Resistance to blast, BPH, BLB and GLH
O. officinalis	24	CC	Tropical and subtropical Asia, Tropical Australia	Resistance to thrips, BPH, GLH, WBPH, BLB, stem rot and tolerance to heat
O. eichingeri	24	CC	South Asia and East Africa	Resistance to BPH, WBPH and GLH
O. latifolia	48	CCDD	South and Central America	Resistance to BPH, BLB, high biomass production
O. alata	48	CCDD	South and Central America	Resistance to striped stem borer and high biomass production
O. grandiglumis	48	CCDD	South and Central America	High biomass production
O. australiensis	24	EE	Tropical Australia	Resistance to BPH, BLB and blast, drought avoidance , tolerance to heat and drought
O. schwein-furthiana	48	BBCC	Tropical Africa	
Meyeriana complex				
O. granulata	24	GG	South and South Asia	Shade tolerance & adaptation to aerobic soils
O. meyeriana	24	GG	Southeast Asia	Shade tolerance & adaptation to aerobic soils
Ridleyi complex				
O. longiglumis	48	HHJJ	Indonesia and New guinea	Resistance to blast and BLB
O. ridleyi	48	HHJJ	South Asia	Resistance to tungro virus, stemborer, blast and BLB
Unclassified (Not belongs to any complex)				
O. brachyantha	24	FF	Africa	Resistance to Yellow stem borer & leaf folder and tolerance to laterite soil
O. schlechteri	48	HHKK	New guinea	Stoloniferous

BPH: Brown plant hopper, GLH: Green leaf hopper, WBPH: Whitebacked plant hopper, BLB: Bacterial leaf blight, CMS: Cytoplasmic male sterility.

The subspecies or varietal groups of *O. sativa* viz., indica, japonica and javanica are the result of centuries of selection by man and nature for desired quality and adaptation to new niches. These can be distinguished on few key characteristics such as glume size, number of secondary panicle branches (rachii), panicle thickness etc.

India Type: It is tropical type rice. It is extensively grown in India, Srilankha, Thailand, Malaysia and adjacent countries. These are tall plants, weak stem, photoperiod sensitivity, easy shattering, and broad droopy leaves and having grain dormancy.

Japonica Type: These are temperate type extensively grown Yangtze river valley of china, Korea, and Japan, these are tolerant low temperature, have more leaves, few tillers, relatively resistant to shattering. The grains are short, broad low in amylase content. Rice becomes sticky on cooking.

Javanica Type: These are intermediate type. These are cultivated in Indonesia, Phillippines, Taiwan and Japan. These are tall, thick stemmed, low to tillering and resistant to shattering. These varieties have broad stiff leaves, long awns and large bold grains.

Floral biology:

Inflorescence: A terminal panicle; Borne on top most internode. The panicle may be close or open, compact or loose, erect or drooping. The main rachis is angular which bears primary and secondary branches. It is branched with a number of spikelets attached to small rachillae. Spikelets are pedicelled and occur on secondary branches. They are solitary at the end. In some varieties spikelets occur in clusters.The rachilla bearing the spikelet has a little bent, above which an enlarged annular portion is, on this two lower glumes are placed.

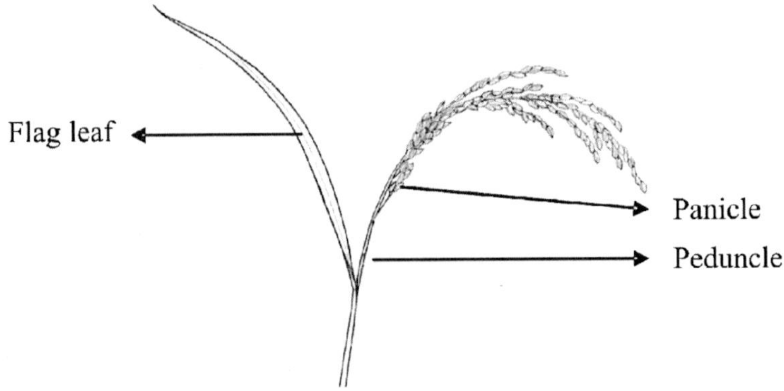

Spikelet: Each spikelet is one flowered and laterally compressed. Each Spikelet consists of two glumes and one lemma and palea. Glumes are small, lanceolate, coriaceous and shiny. Glumes are 1/8th (one eighth) of the size of the spikelet. When the glumes-are bigger than the spikelet (More than half) they are said to be winged spikelet. "They play an important role in keeping lemma and palea in position during pollination. Lemma is boat shaped, 5-nerved. In some varieties die tip of the lemma may prolong into awns. Palea is boat shaped, 3-nerved. The tip of the lemma together with the tip of the palea constitutes, "Apiculus". The perianth is represented by two fleshy thick lodicules, which are hygroscopic and help in the opening of the spikelet.

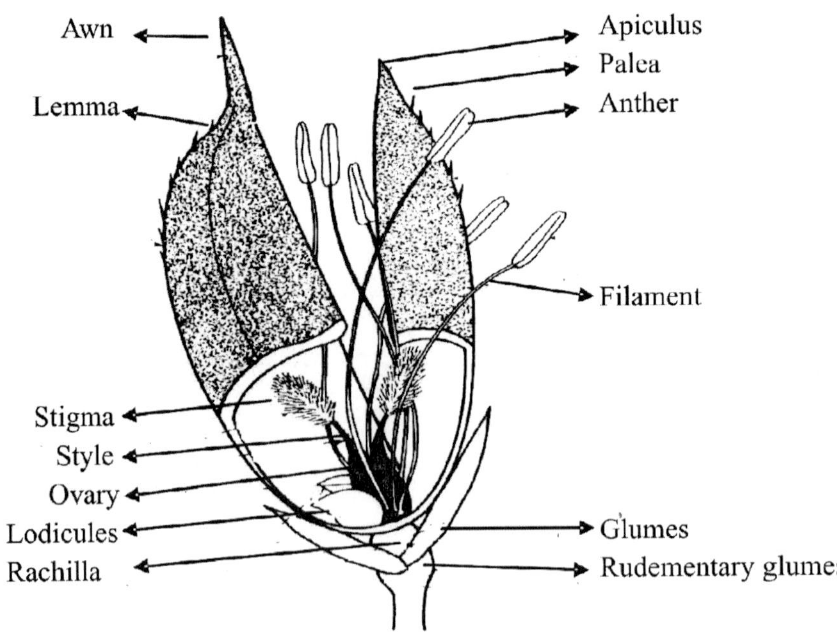

Androecium consists of six stamens present in two whorls of three each. Anthers are dithecous, introrse and versatile. Gynoecium is monocarpellary, unilocular, superior ovary with one ovule on basal placentation. Styles are two each with a plumose, bifid, feathery stigma. Fruit is a caryopsis, generally called grain enclosed between lemma and palea. Both lemma and palea constitute the hull. The fruit wall and seed coat are fused.

Floral Formula: Br, Brl, %, ⚥, P_2, A_{3+3}, $\underline{G}_{(1)}$

Floral Diagram:

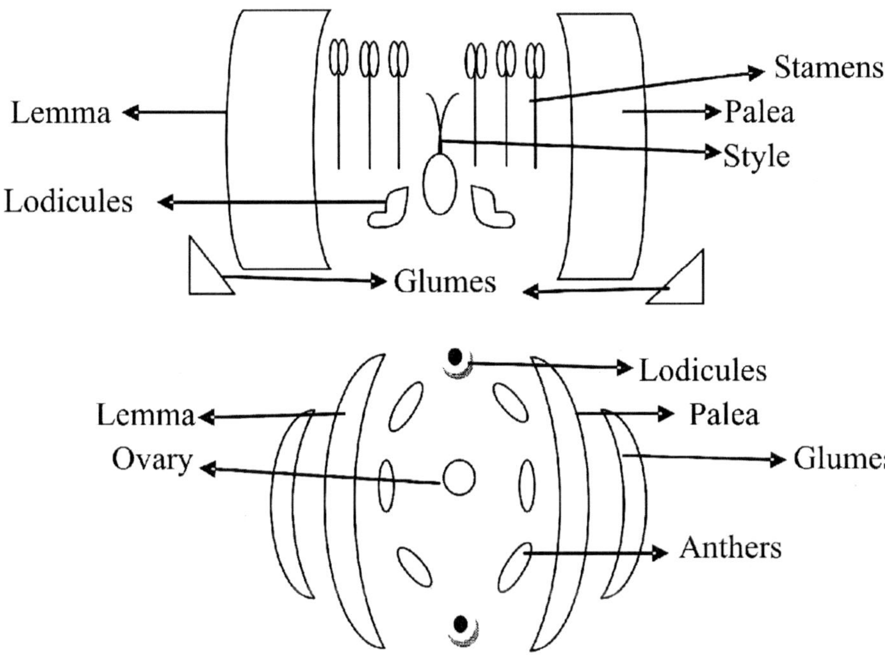

Anthesis and pollination: Rice is a protandrous crop and self pollination takes place. Panicle takes 2 to 3 days for complete emergence from the boot leaf. Maximum flowers open 2nd and 3rd day after the complete emergence. The top most spikelet opens first and flowering proceeds from top to downword. But in branches flowering proceeds upwards. On main axis there will be 8-10 nodes giving rise to 8-10 primary branches on which are present secondary and tertiary branches. The blooming of rice normally occurs between 10 A.M. and 2 P.M. The time and rate of blooming varies with the cultivar and is affected by environmental factors such as temperature, humidity and light. It is delayed during cloudy days. The stigma is receptive for first

3 days after opening of the spikelet. Pollen is viable for a very short time i.e., only 5 minutes. Rice pollen can be preserved for 24 hours at 54°F at 95% relative humidity. Extent of natural crossing varies from 0 to 3% depending upon variety, season and environment. In rice, selfing is done by covering the panicle with butter paper bag for obtaining 100% pure seeds of the variety. Emasculation is done by the methods; Forced open method, clipping method, hot water treatment, cold water treatment, hot air method and relative humidity method. Clipping method is most common method. In this clip off top 1/3rd portion of spikelet using scissors and removes anthers with forceps. The rice panicles are pollinated the morning following emasculation. Rice flowers may be pollinated by placing ripe anthers in each flower by hand or twirling a panicle shedding pollen over the emasculated panicle. After pollination, the panicles are covered for protection from stray pollen.

Breeding objectives: Major breeding emphasis will continue on improving yield, grain quality, and pest and disease resistance.

1. Yield: Grain yield of rice is a complex character that results from the combination of many morphological traits and physiological processes. These along with interaction with environment decide the yield potential. To get higher yield we must have to develop a ideal plant type with short stature, thick and stiff culm, compact panicle to hold the plant erect, shrot, narrow, erect leaves to effectively utilize solar radiation, high tillering, non/low photosensitivity and nitrogen response. A real breakthrough in yield potential in rice yields came with the development of short statured semi dwarf cultivars under irrigation and intensive management.

2. Quality: The important quality traits includes grain shape and grain size, texture of endosperm and quality of starch, aroma and cooking quality and milling outturn etc. Development of rice with β-carotene pathway and high iron content is also important trait. These are briefly described below in the table.

S. No	Trait	Preferred type	Remarks
1.	Hulling and milling recovery	High	Range,70-79% (Total rice)
2.	Endosperm appearance	Transparent	Opaque appearance due to loose packing of starch, polygenic; low heritability
		Non-waxy	A single dominant gene

3.	Length and shape of grain	Long slender	Five groups :LB, LS, MS, ShB, Shs; polygene; high H
Cooking quality			
1.	Amylase content	Intermediate	Groups: Low (up to 20%),intermediate (21-25%), high (>25%);single gene
2.	Grain elongation on cooking preferred	Preferred	Special trait of 'basmati' rice
3.	Aroma	Presence of aroma	Scented types fetch, higher price; oligogenic; recessive

L-long, S-slender, M-median, B-bold, Sh-short, H-heritability.

3. Resistant to biotic stresses (pests and diseases): Pests and diseases are important constraints to rice production and breeding for pest and disease resistant lines is one of the primary aims of rice improvements programme. Blast, Helminthosporium, Bacterial leaf Blight, Sheath blight and Tungro virus are some of the important diseases. And Brown plant hopper, Stem borer, Rice gall midge are important pests. In addition to race specific resistance induced by major genes, general resistance, which results in the production of fewer and/or smaller lesions, has been found to be more durable line of defense.

4. Resistant to abiotic stresses: Breeding for abiotic stresses like cold temperature, drought, salinity and alkalinity or deficiencies in zinc, iron and phosphorous are important constraints in rice improvement. The japonica races that evolved in china, korea and japan have greater tolerance to cool temperatures than the indica races that evolved in tropical Asia. Cool temperature during pollen formation increases sterility and reduces yield. High temperature may severely restrict pollen shedding in heat-susceptible cultivars. In drought situations, the deeper root growth enables the rice plant to obtain water from deeper soil levels and maintain high leaf water potential.

5. Resistance to herbicide was proposed to be particularly important for upland rice and for allowing direct seeding to replace the practice of large scale transplantation.

6. Resistant to lodging and shattering: Varieties with short stature with strong culm and short internode are resistant to lodging and even capable of supporting their own body even if high grain yielding trait is introduced. The short stature was obtained by the introduction of dwarfing gene "Dee-gee-woo-gen". In rice grain shattering is mainly caused by seed abscission.

7. Breeding for early maturity: This is mainly desired for multiple cropping and also helpful to escape from terminal drought, pests and diseases. Additionally, early maturity rice utilizes less irrigation water or permits harvest of a ratoon crop after the initial crop.
8. Breeding for wide adaptability: Varieties with insensitive to temperature and light intensity are wider adaptable over a wide range of agro climatic regions.

Breeding methods

1. Introduction: The variety Taichung native-1 (T-N-1) is the cross between

   Dee-geo-woo-gene and Tsai Taichung is imported in 1960 for seed multiplication and distribution. Another variety IR-8, a tall and heavy tillering disease resistant developed from cross between Indica variety of Indonesia and dee-geo-gene imported in 1966.
2. Selection (pureline): From land races outstanding lines are selected and evaluated along with existing varieties and superior types are multiplied and released for commercial cultivation. The varieties developed by selection are Terna, Tuljapur-1, Ambika, Ambemohor pussa -33, etc.
3. Hybridization and selection:

   Pedigree method: It is the most common method of rice breeding. Crosses made to combine genes for desirable characteristics from two or more parents. Seeds of such crosses sown in pedigree nursery. Selections are made in F3 and onwards generations and superior progenies are carried forward and tested with existing check variety and if found superior released for commercial cultivation. This includes inter varietal, inter racial e.g. Japonica × indica cross ADT 27 (Norin 10 × GEB 24) and ponni (mashuri) (Taichung 65 × ME 80) and inter specific crosses e.g. Co 31 (*O. perennis* × GEB 24). Bulk selection method is not followed in rice but single seed decent known as rapid generation advance is in use in IARI. Two to three generation in green house can be grown in a year.
4. Mutation breeding. It includes spontaneous mutation (GEB 24 from Athur Kichili Samba known as KONAMANI, fine Grain and Quality Rice) as well as induced mutations (Jagannath rice from Orissa).
5. Heterosis breeding: Rice hybrids e.g. APHR-1 (IR 58025A × Vajram), APHR-2 (IR 62829A × MTU9992), KRH1 (IR 58025A × IR 9761-10-IR) and CORH-1 (IR 62829A × IR 10198 - 66-2 R) etc.

6. Transgenic breeding: Resistance sources for majority of stresses were thought to be inadequate, therefore, the development of transgenics for resistance to the biotic and abiotic stresses needs to be given the highest priority.

Hybrid rice: Heterosis in rice was reported by Jones in USA as early in 1926 and Ramaiah in 1933. But the hybrid rice technology was initiated in 1964 in China by Yuan Long Ping (Father of hybrid Rice). However, only in 1970, when a wild abortive pollen plant was identified in Southern China, did the idea begin to materialize. In 1980, Shih-Cheng and Loung Ping (1980) published one of the first articles indicating the potential of hybrid rice. The proposed strategy then relied on the male sterility produced by the abortive pollen system identified in the wild species O. sativa L. f. spontanea.

Breeding techniques involved in development of hybrid rice are three line system of breeding, where one line would have the genetic cytoplasmic male sterility; the second line would be responsible for maintaining the sterility, and a third one would be used as the matching parent for the hybrid with the responsibility of restoring the fertility. Two line system of hybrid breeding evolved through Environmental induced genetic male sterility (EGMS) through either photosensitive genetic male sterility (PGMS) or thermosensitive genetic male sterility (TGMS) which had been used successfully in china and also by application of gametocides. One line system of hybrid breeding evolved through apomictic lines or anther cultured hybrids.

List of some released varieties/hybrids in rice in India

Name of variety	Parentage	Salient Features
TaraniBasmati	From local basmati	Tall grains-slender, scented yield-21Q/Ha
Samba mahsuri (BPT 5204)	(GEB 24 × T(N)1) × Mahsuri	Fine grain (medium slender) and good cooking quality. Recommended for irrigated uplands.
Sona mahsuri (BPT 3291)	Sona × Mahsuri	Fine grain, dwarf and high yielding.
Erramallelu (WGL 20471)	BC 5-55 × W 12708	Highly resistant to gall midge (biotype 1).
Ranbir Basmati (IET11348)	Basmati370-90-95	Grains - long slender scented
Tellahamsa	HR 12 × T(N)1	Wide adaptability, long slender grain type. And cold tolerant.

Pusa 677	Pusa 312 × TKM-9	Semi dwarf ,grains medium, slender resist to neck blast & leaf scald. More resistant to leaf blast
Mahsuri	Taichung 65 × Myang Ebos 80/2	Fine grain, tall productive even under low inputs and low intensity of sun light.
Swarna (MTU 7029)	Vasishta × Mahsuri	Drak green foliage, high yielding variety comes up very well even under low nitrogen input.
Jagital sannalu (JGL 1798)	Samba mahsuri × Kavya	Non lodging, non shattering, medium slender grain, resistant to gall midge.
Polasa prabha (JGL 384)	Samba mahsuri × kavya	Non lodging, non shattering, medium slender grain, fertilizer responsive culture. Tolerant to BLB, BPH, blast and gall midge.
Nandyal sannalu (NDLR 8)	BPT 3291 × CR 157-212	Suitable for late planting conditions, tolerant to BPH, leaf folder and moisture stress.
Sumathi (RNR 18833)	Chandan × Pak-basmathi	Semi tall scented, non shattering culture. Photo insensitive moderately tolerant to drought, tolerant to blast and gall midge.
Pusa 834	Pusa 312 × TKM-9	Grains long & bold
Vaidehi	TCA-48	Tall grains, long and bold, resistant to blast BLB. Tolerant to draught. Yield 30-35Q/Ha
ASD-19 (IET-104360)	Nalkanda × IR 30	Semidwarf, grains slender, white, resistant to blast
ADT-42	AD 9246×ADT 29	Semidwarf grains, long and slender white resistant to blast
TPS-3	RP-31 × 49-2X (LMN)	Semidwarf, grains long slender and resistant to blight
Poornima (IET-12284)	Poorva × IR 8008-298	Semidwarf, resistant to major pest and diseases
Indur samba	BPT5204×SUREKHA	Medium grains, tolerant to GM
Radhi	Swarnaprabha mutant	Mutation breeding
Jalprabha	-	Composite selection
Satyaranjan	IET 9711 × IET 11162	Tall grains medium slender, white, resistant to blast
Pavithra	Surekha × Mo5	Grains medium bold

Remanica	M01 Mutant	Short, bold grains
Revathy	Culture 12814 × MO-6	Grains medium, bold resistant to BPH
TKM - 11	-	Semidwarf grains– medium, bold, white, resistant to blast and tolerant to BPH
ADT -43	-	Dwarf, grains medium slender, white resistant to GLH, BPH, GM
WGL -3962 Bhadrakali	-	Grains medium, slender, resistant to GM
ADT RH -1	-	Grains– long slender, yield-71 Q/ha
MDU -5	-	Dwarf, grains, medium, slender white, resistant to drought suitable for dry land transplanted condition
Triguna	-	Grains slender, long tolerant to BPH and GM
Triguna (IET – 12785)	Swarnadhan × RP 1570-38	Grains long slender, tolerant to BPH & GM
IET -8116	Vikram × Andrewsali	Grains long bold
Krishna Hamsa (IET -9219)	Rasi × fine gora	Semidwarf , grains long slender , long fine ; yield 50-56 Q/ha
Vijetha (IET – 13967)	MTU 5249 × MTU 7014	Semi dwarf grains , medium, slender AWA tolerant to BPH X Blast
COR – 46 (TNAU BPHR -8)		Tall grains – long slender AWA resistant to BPH, yield 60 Q/ ha
ASD -20(IET-13181)	IR 18348-3 × IR 25863 61– 3-2 X IR-58	Semidwarf grains , long slender , white resistant to SB, LF and resistant to blast and yield 67 Q/ ha
PHB -71	RF 1301 × RM 1401	Semi dwarf grains long slender , tolerant to blast, BPH, GM
Sravani	Selection from IR 50	Dwarf (80-85) cm, grains- long slender, white, resistant to blast
Somasila	Selection from IR 50	Resistance to blast, helminthosporium and tolerant to BLB.
Swathi	IR-36 ×MTU-4569	Susceptible to RTV, SB, LF, RH, Rice thrips and BPH.
Vedagiri	NLR9672-96×IET7230	Semi dwarf (107cm), tolerant to SB, GM and RTV, Susceptible to BPH, LR.
ADT-44 (IET 14099)	Selection from OR-1126-7-1	Semi dwarf, (112 cm), grains short bold, white, resistant to blast

CO-47(IET-14298)	IR50×CO-43	Grains medium, slender, resistant to blast, moderately susceptible to hoppers.
Early Samba (RNRM-7)	Mutant of BPT 5204	Dwarf (80-90 cm), grains medium , slender, tolerant to SB.
Deepthi (wnd-111)	Pure line selection from edavaka	Tall 123 cm, grains long, slender, red, moderately resistant to blast.
Hybrid-6201 (pa-103)	CO-2×MO-1	Grains long, bold, resistant to blast, tolerant to SB, BPH.
Surya(BPT-4358)	Sona mahsuri× ARC-6650	Semi dwarf, grains long, slender, white, tolerant to BPH.
KAUM-20-19-4 (IET-13981)	Mutant to MO-1	Dwarf (90-93cm), grains medium, bold, red, resistance to BPH.
Makaram(KTR-2)	-	Semi dwarf, grains medium bold, red.
Kumbhan(KTR-3)	-	Semi dwarf (95 cm), tolerant to lodging , no major disease and pest.
APRH -1	IR – 58025A × Vajram (MTU 5249 (R)	Semi dwarf grains long slender white, AWA, yields-35Q/ha
APRH -2	IR-62829 × MTU 9992 (R)	Semi dwarf grains, Long slender, white, yield-45Q/ha
Karnataka Hybrid rice	IR 58025A × IR9671R	Yield-60 Q/Ha
Karnataka rice hybrid -2(KRH-2)	IR-58025A×KMR-3R	Semi dwarf 102 cm, white, slender, tolerant to LB, BS and other diseases.
DRRH -1	IR58025 A × KMR 3A	Grains– long slender
Karnataka rice HB -2 (KRH -2)	IR58025A × KMR3A	Grains long, bold

Economic importance: It is mainly used as a food crop (parched rice, rice flakes and puffed rice). Rice wine, which contains 10-15 percent alcohol, is usually made from glutinous rice. The husks are value less as an animal feed and little value as fertilizer. They are used as fuel for rice mills, as an aggregate for light weight concrete and for making hard board. Rice bran is valuable livestock and poultry feed. The rice starw is used for cattle feeding, rice bran oil is used in cooking.

Wheat

Wheat (*Triticum sps.*) is the most important food grain of the temperate zones, giving about one-third of the total production, followed closely by rice. It has been the basic staple food of the major civilizations of Europe, West Asia and North Africa. It mainly distributed in countries like USA, Canada, Latin America, Europe, China, Japan, Argentina, Mexico, India, and Pakistan. Every month of the year a crop of wheat is harvested somewhere in the world. Today, wheat is grown on more land area than any other commercial crop and continues to be the most important food grain source for humans. Its production leads all crops, including rice, maize and potatoes. Wheat mainly belongs to the genus *Triticum* family poaceae. Centre of origin for diploid species is Asia minor, for tetraploid is Abyssinia, North Africa and for hexaploid it is CentralAsia.

Species of the genus *Triticum:* According to vavilov genus consists of 14 species. Which includes *T.boeoticum, T.monococcum, T.dicoccoides, T.dicoccum, T.durum, T.persicum, T.turgidum, T.polonicum, T.timopheevi, T.aestivum, T.sphaerococcum, T.compactum, T.spelta* and *T.macha*. Of which some are diploid, some tetraploid and some hexaploid species.

Species	Genome	Distribution	Remarks
Diploid (2n=14)			
T.boeoticum (Wild einkorn)	AA	S.E. Europe, W. Asia, Transcaucasia	Wild ancestor of *T.monococcum*
T.monococcum (Einkorn)	AA	Locally grown	Widely cultivated in ancient times
Tetraploid (2n=28)			
T.dicoccoides (wild emmer wheat)	AABB	Syro-palestinian region	Wild ancestor of most tetraploid wheats and amphidiplod form resulting from the hybridization of *T.boeoticum* and *Ae.speltoides.*
T.dicoccum (emmer wheat)	AABB	Locally grown	Widely cultivated in ancient times and oldest of the cultivated wheat
T.durum (Macaroni wheat)	AABB	Mediterranean	Cutivated and Grains contain high glutin.
T.persicum (Persian wheat)	AABB	Iran, Iraq, Caucasis	-
T.turgidum (Rivet wheat)	AABB	Mediterranean	-

T.polonicum (Polish wheat)		S. Europe, Near east, India	-
T.timopheevi (Timopheevi wheat)		Georgia	-
Hexaploid wheat (2n=42)			
T.aestivum (Bread wheat)	AABBDD	Temperate and continental regions	Cultivated
T.sphaerococcum (Dwarf wheat)	AABBDD	India	-
T.compactum (Club wheat)	AABBDD	Temperate and continental regions	-
T.spelta (Spelt wheat)	AABBDD	Iran, Europe	Amphidiploid from hybridization between *T.dicoccoides* and *Ae.squarosa.*
T.macha (Makha wheat)	AABBDD	Georgia	-
Species other than *Triticum* (Diploid 2n=14)			
Aegilops speltoides	BB	-	Wild and donor of the B genome
Triticum dichasians (Aegilops caudata)	CC	-	Wild
Triticum tauschii (Aegilops squarrosa)	DD	-	Wild and source of D genome

Evolution of the cultivated wheat species: The tetraploid species developed as an amphidiploid between two diploid species, and hexaploid species originated from a cross between tetraploid and diploid species. It is believed that tetraploid species *Triticum turgidum* evolved as an amphidiploids between *Triticum monococcum* (AA) and unknown species (now probably *Aegilops speltoids*) with BB genome. *Triticum monococcum* was originated from *Triticum boeoticum* with natural mutation and selection. So *Triticum boeoticum* was probably considered as the ancester for all the cultivated wheats. The hexaploid bread wheat originated as an amphidiploids between *Triticum turgidum* and *Triticum tauschii*. The overall process represented as follows.

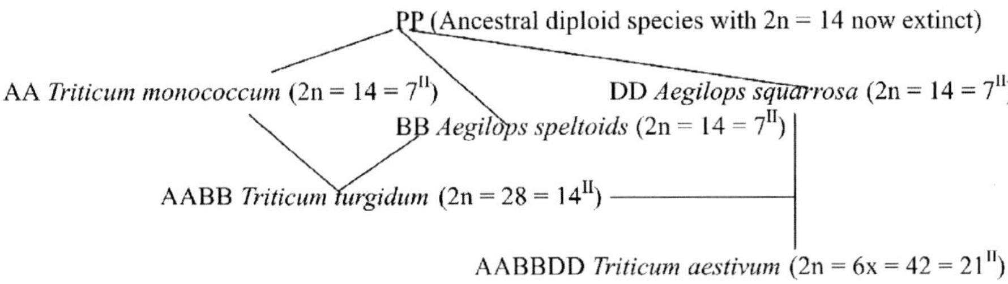

The 21 chromosomes (gametic number) of hexaploid wheat have been divided into seven groups, called *homeologous groups*. Each homeologous group contains three partially homologous chromosomes, one chromosome from each of the A, B and D genomes. The three chromosomes within the ABD homeologous group frequently contain loci in common for a particular character. For example, two genes for leaf rust resistance are located on chromosome 2A, three genes on chromosome 2B, and three genes on chromosome 2D.

Floral biology: The inflorescence is terminal spike or head bearing 15-20 spikelets. Which are arranged singly at each notch of the zig-zag rachis, in the shallow depression on the rachis. The rachis is broader above and tapering below convex towards outerside and concave towards innerside. Spikelets are arranged in two rows. The top most spikelet is often sterile.

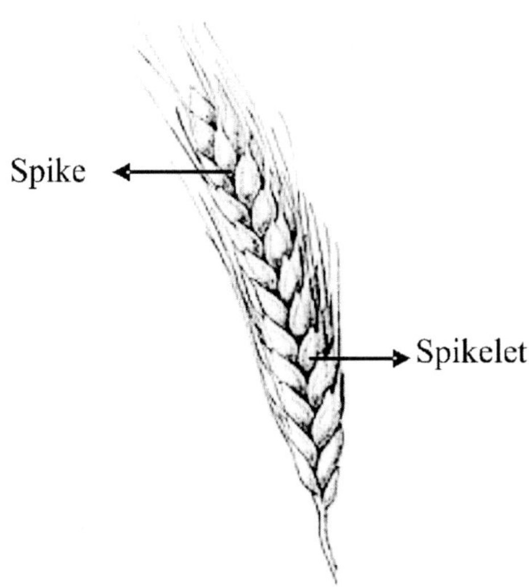

Spikelet: It consists of two glumes and three or more florets. Glumes are boat shaped. Each glume has round base ending in an acute tip or a long bristle or awn. Each floret is having a lemma, which is boat shaped and possesses a very long awn, palea is membranous, hyaline and slipper shaped, complete, hypogynous, regular, zygomorphic, bisexual, cyclic, homochlamydeous, perianth is represented by two lodicules, which are cuneate in shape and slightly fleshy. Only lateral florets are fertile and the central florets may be either sterile or staminate.

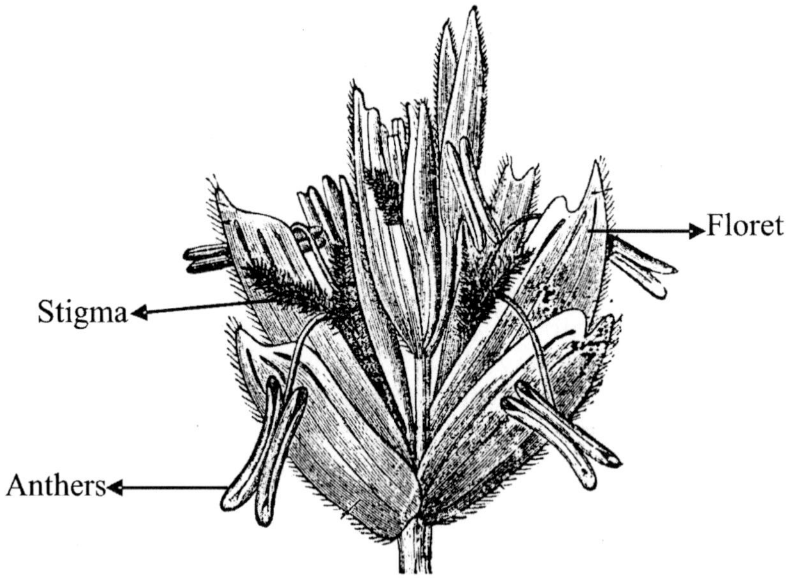

Androecium consists of three stamens with thin filament and large anther. Anthers are dithecous, introrse and versatile. Gynoecium is monocarpellary, unilocular, superior ovary with one ovule on basal placentation, styles two each with a plumose, feathery stigma. Fruit is a caryopsis, oval in shape and a tuft of hairs present on the stigmatic end called as brush. The dorsal surface is smooth and round. The ventral surface has a longitudinal groove called as crease, which indicates the line of attachment of ovule to placenta. The either sides of the crease are called cheeks. Embryo is situated at the base on dorsal side.

Floral formula: Br, Brl, %, , P_2, A_3, $\underline{G}_{(1)}$ (For lateral florets)

Br, Brl, %, P_2 (For central florets)

Floral diagram:

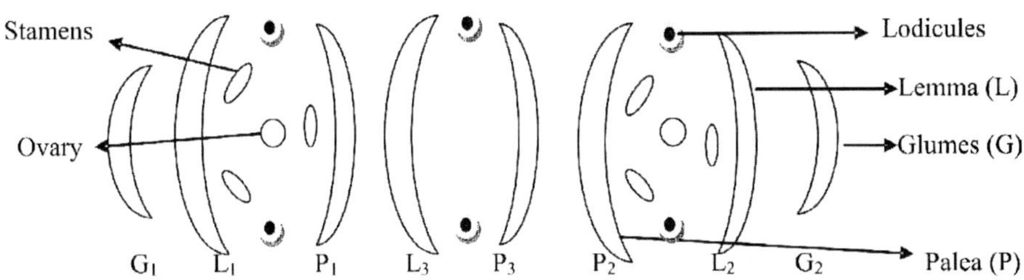

Anthesis and Pollination: The spike takes two days for complete emergence from the boot leaf. Flowering starts at 2/3rd height of the spike and proceeds in both the directions. Flowering in wheat occurs throughout the day maximum in between 10.00AM to 01.00 P.M. Lodicules helps in exertion of lemma and palea. In a spike blooming is completed in 2-3 days. Anthers and stigmas come out simultaneously before the emergence of anther dehisce and shed pollen on the stigma resulting in self pollination. While the flower is open, foreign pollen may enter, normally resulting in about one to two percent cross pollination. If conditions are unfavourable for the opening of the glumes, the anthers may shed the pollen without being extruded. Selfing is done by bagging the inflorescence of the desired parent with parchment paper bag before commencement of anthesis. For emasculation, the upper and lower spikelets are clipped off by maintaining about 15-20 spikelets per spike. In each spikelet only the lateral fertile florets are kept, all others are removed. The main emasculation methods used in wheat are clipping method, forced open method and chilling method. In wheat, commonly flowers are emasculated in preparation for crossing by clipping back the glumes, removing the anthers with fine-pointed tweezers before pollen is shed, and covering the wheat spike with plastic bag to protect the flower from foreign pollen. Pollinations are made one to two days later by applying the ripe anther over the stigma. Pollen remains viable for one to three minutes and fresh pollen is essential for good seed set.

Breeding objectives: Major breeding emphasis will continue on improving yield, grain quality, and pest and disease resistance.

1. Yield: In wheat, achievement of high yield mainly depends on the improvement three components which include number of heads per unit area, number of grains per head and the average weight of the grain. Although many varieties in the past have been designated as high yielding when compared to the cultivars replaced recently are described as high yielding have resulted from the exploitation of short statured cultivars with dwarfing genes i.e. semidwarf varieties.

2. Quality: In wheat, the main quality trait is gluten content. Gluten is an indefinite, tough, nitrogenous compound that remains after wheat flour is washed to remove the starch. The soft wheat contain lesser gluten which is suitable for making confectionary products such as cakes, cookies, pastries, crackers or flat bread etc. While hard wheat or bread wheat which has higher gluten content. Durum wheats have the greatest hardiness and are milled to produce a coarse, granular product called semolina, used in making pasta, and durum flour used for making noodles.The important quality traits are given in the below table.

Trait	Preferred type	Remarks
Grain size	Bold, pump	-
Grain colour	Amber to white	-
Grain texture	Hard, vitreous, lustrous	-
Protein content	high	Atlas 66 and Naphal have over 19% protein
Gluten content	High	(C 273 and UP302 have high gluten (13%)
Polshanke value (Gluten strength)	Strong (>150 min) Medium strong (81-150 min) Weak (<80 min)	Suitable for bread making Suitable for chapathi making Suitable for biscuits and confectionery
Sedimentary value	(20-40)	All purpose wheat; lower the value, weaker the gluten
Alveogram	Strong (>300),medium strong (200-300)), weak (< 200)	Measures gluten strength
Mixogram	-	Indicates baking quality also glutens strength

3. Resistance to biotic stresses: Breeding for resistance to fungal diseases appeared as the most important overall objectives in present day wheat breeding programmes. Leaf rust, stem rust, leaf blight, kernel bunt, stripe rust and loose smut was identified as major diseases in wheat. Most virulent stem rust race is ug99. There were resistance genes Sr24, Sr31, Sr36 for stem rust, Lr1to Lr34 for leaf rust and Yr6, Yr7, Yr9 and Yr29 for stripe rust identified. So breeding for rust resistance is mainly focused to improve the horizontal resistance through the development of multilines. Among the insects Hessian fly, green bugs and cereal leaf beetle are the major pests. This we can achieve through the Antibiosis mechanism.

4. Resistance to abiotic stresses: there is need to breed varieties with tolerance to drought, heat, water logging, cold, salinity etc. plant mechanisms that contribute to the drought resistance are early maturity to ripen the crop before the period of drought stress, vigourous and deep root system to utilize available soil moisture efficiency, ability to close stomata during periods of drought stress to decrease water loss, and a waxy bloom on the leaf surface to reduce transpiration loss.
5. Resistance to lodging: breeding for resistance to lodging involves development of cultivars with a vigorous and healthy root system, short stature, resilient straw that does not break in the wind and resistance to diseases and pests for attack and weaken the straw or the root system. The short stature was obtained by the introduction of dwarfing gene "Norin-10". Genes that reduce height in wheat are given the symbol "Rht". This is achieved after the identification of dwarfing gene "Norin 10" E.g. Sonara 63, Sonara 64 and Kalyan sona.
6. Winter hardiness, early maturity and yield stability are some of the important breeding objectives.

Breeding methods

1. Introduction: Semi dwarf wheat varieties Sonara 63, Sonara 64, Mayo 64 and Lerma Roja 64 were introduced from Mexico.
2. Selection (Pure line): Earlier varieties like Pusa4 and Pusa12 (now New Pusa4 and New Pusa12) were developed at IARI as a result of pure line selection. NP4 was adjusted asone of the best quality times during 1916 to 1920 in international grain exhibition.
3. Hybridization and selection: Inter varietal, varieties like Kalyan Sona, Safed Lerma and Sharbhati Sonara were released. Inter specific crosses especially back cross method of breeding is the target to achieve the pest and disease resistance.
4. Mutation breeding: Sharbathi sonara with increased protein content was released.
5. Development of multilines: N.E. Borlaug (1953) developed multilines against rust. Multiline is a mixture of isogenic lines. Each line is having resistance against a particular race of a disease. And each line is produced by separate back cross method of breeding. In India, three multiline varieties, KSML 3, MLKS 11 and KML 7404 have been released in wheat from Punjab. N. E. Borlaug (1965) believed that the development of a multilineal hybrid variety will

offer the best possible means of incorporating a long lasting rust resistance in to a commercial wheat variety. Once an outstanding single-cross hybrid has been developed, it can be converted into a multilineal hybrid.

a. The first step in this process is the development of many lines phenotypically identical with the male sterile female variety A^S (the modified variety of A) through a back-crossing programme.

b. Each of these lines, designated male variety A^n, will be genotypically different from the other in stem rust resistance, and also from that of the male sterile female variety A^S, but none of these lines will possess the genes for restoration of fertility.

c. Therefore, when crossed with the male sterile female variety A^S, the resulting F_1 plants of each cross will be sterile.

d. These male sterile F_1 lines in turn each will be crossed as female with the male variety B^r to provide a heterotic effect, and this second cross will also simultaneously restore fertility to the F_1 commercial hybrid seed formed in the process.

e. The basic commercial variety A and B, both of which are fertile and into which the cytoplasmic sterile and fertility will be incorporated respectively, should also have stem rust resistance gene.

f. Each individual F_1 plant of a hybrid produced by this method would be different in its resistance to stem rust and rust epidemics is not likely to develop in such a population.

Cross 1 (Primarily for rust resistance) — Cross 2 (Primarily for heterosis)

Variety A^S:
× ♂ Var. A^n (Stem rust resistance 1) → Sterile ♀ F_1s × ♂ Var. B^r
× ♂ Var. A^n (Stem rust resistance 2) → Sterile ♀ F_1s × ♂ Var. B^r
× ♂ Var. A^n (Stem rust resistance 3) → Sterile ♀ F_1s × ♂ Var. B^r
× ♂ Var. A^n (Stem rust resistance 4) → Sterile ♀ F_1s × ♂ Var. B^r
× ♂ Var. A^n (Stem rust resistance 5) → Sterile ♀ F_1s × ♂ Var. B^r
× ♂ Var. A^n (Stem rust resistance 6) → Sterile ♀ F_1s × ♂ Var. B^r
} Multilineal Hybrid Variety

Scheme for development of multilineal hybrid wheat variety

6. Transgenic breeding: overexpression of triticin gene for improving lysine content was identified as one of the important objectives of transgenic research. Engineering of β – carotene pathway in wheat would be useful for addressing the problem of malnutrition. Development of transgenics for resistance to diseases and pests needs to be given the priority.

Hybrid wheat: At Kansas Agri. Expt. Station USA male sterile lines were identified by crossing *T.timophevi* × *T. aestivum*: Bison variety. By repeated back crossing a male sterile line resembling Bison was evolved. The male sterility is resulted from the interaction of nuclear genes from *T. aestivum* with cytoplasm of *T.timophevi* and fertility is restored by genes from *T.timophevi* and other sources. Utilization of the system as proposed for the production of hybrid wheat involves the development and maintenance of cytoplasmic male sterile A-lines and their maintainer B-lines, development and maintenance of fertility restoring R-lines and crossing A-lines with R-lines for the commercial production of hybrid wheat. At present USA and Canada are doing work on this. Commercially hybrid wheat seed has been produced using chemical hybridizing agents, i.e., plant growth regulators that selectively interfere with pollen development, or naturally occurring cytoplasmic male sterility systems. XYZ system of male sterility for production of hybrids in wheat was proposed by Driscoll in 1972. X = Disomic addition line 2n + 1IIR wheat genome carry gene for male sterility Rye chromosome has gene for fertility restorer. This is closely linked with marker hairy peduncle. They are male fertile. Y = This component has only one dose of alien rye chromosome so that it produces two types of male gamete (with alien - non functional; without alien- functional)

Similarly two kinds of female gamete are produced: [(i) with alien chromosome which on fertilization give rise to a 'Y' plant and (ii) Without alien, which is non kairy and is sterile].

Z plant = 2n of wheat = male sterile

Y-plant = (2n + 1R)

(Male fertile)

Selfed

2n (Z-plant) | 2n + 1R (Y-plants)

X-plant (2n+2R)

(Male fertile)

↓

On selfing

↓

X-plant (2n+2R) × Z-plant (2n)

↓

(2n+1R) Y-plant

Z-plant are male sterile, it bear the hybrid seed. X and Y needed for the production of female parent of the hybrid.

List of some released varieties/hybrids in wheat in India

Varieties	Parentage	Special traits
PBW-373	-	High yeilding & disease resistant, vareity possesing late heat tolerance.
PBW-343	-	Wide adaptability, resistance to rusts &tolerance to karnal bunt.
HD-4672 Malav patna	Bijaga red × PDW34	Medium late maturity & white at maturity. Fars glabrous &black.
LOK-45	S308/S331	Intermediates growth habit,early maturity, amber, semi hard.
GW-1189	SU-gullugu-7 × VD90-12	-
HI-1500(amrita)	HW 2002 2	Semi erect type, day to maturity amber
NIDW-295 (Godavari)	BOOMER33	-
UP-2565	PBW352	-
HD-2833(TRIPTI)	PBW226 × HW1042	-
VL-829	IBWSN	Semi spreading type, late maturing, amber semi hard & bold grains.
HS-420(Skivalik)	LAJ 3302	Erect type, amber, semi hard & smaller grains.
HD 5001	-	High degree of resiarance to leaf stripe rusts
HS-375(Hingiri)	BT 2549 × FATH	-
DBW-14	RAJ3765	Early maturity, hard grains, amber.
MP-4010	Angostura 88	Early maturity,amber hard & medium grains.
PBW-52W	PBW175	High degree of resistance of leaf stripe rusts.
VL- 832	PBW65	-
SKW-196V (Snalimar wheat -7)	W 485 × PBW 343	-

PDW-291 (Durum)	Boomer 21	High degree of resiatant to loose smut, karnal bunt,rust, falg smut.
PBW- 527V	PBW 343	High degree of resistance to stripe & leaf rust
HD-2864 (URJA)	DL 509-2	Resistance to leaf & stripe rusts, heat tolerance
DDK 1025	DDK 1013	-
HI-8627 (malav kirti)	HD 4672	High degree of resistance to leaf &stripe rusts.
HS-490 (Pusa baker)	HS 364	Resistant against leaf & stripe rust with seedling resistant.
PBW-590	WH 594	High yielding, leaf rust resistant, high protein content.
RAJ 4120	PBW 343	Early maturing variety, highest yeild, better resistance to rusts.
MP-1203	FASN × 28 TEPOKA	Resistant brown &black rusts highest protein.
TL- 1210(triticale	Cinnamon	-
PBW-373	ND × VG9144	High yielding and disease resistance
UP-2382	CPAN 2004	-
WH 711	ALD'S' HUAC	Late maturity, amber, hard grains.
WH 912	HUI'S' × YAV'S'	Amber, hard & bold grains
PDW -274(durum)	DWL 60/8	-

Economic importance: It is used as a food crop.

Maize

Maize is one of the most important cereals of the world and it ranks only next to rice and wheat. Over 80% of maize production is located in the Americas (53%) and Asia (28%), followed by Europe (15%). Key areas of cultivation include the US maize belt, north eastern China and Eastern Europe followed by India, Mexico, Argentina, Ukraine, Indonesia, France and Canada. The genus to which it belongs is *Zea*, tribe Maydeae of the family Gramineae (Poaceae). Center of origin of maize is Central America.

Species of the genus *Zea*: The species *Zea mays* includes both domesticated maize (ssp. *mays*) and its closest wild relatives known as the teosintes. Outside the Zea genus, the closest wild relatives are from the genus *Tripsacum*. Information about wild relatives teosintes and tripsacum, are presented below:

Teosintes: The teosintes are wild grasses native to Mexico and Central America and have limited distribution and are restricted to North, Central and South America. They are not reported to occur in Southeast Asia.

a. The nearest teosinte relative to *Z. mays* is *Z. mays* sp. *mexicana* previously classified as *Euchlaena mexicana, Zea Mexicana*. This is mainly distributed in Central Mexican region and it is a annual teosinte.

b. *Z. diploperennis* and *Z. perennis* are diploid and tetraploid perennial teosintes, respectively, with narrow distributions in Jalisco, Mexico. *Zea diploperennis* was on the threshold of extinction when found in the late 1970s and has since been used extensively for investigating maize ancestry.

c. *Zea perennis* is reported to be established in South Carolina, USA.

d. *Zea luxurians* is an annual teosinte from southeastern Guatemala.

e. *Zea mays* ssp. *parviglumis* is mostly wild teosinte of southern and western Mexico.

f. *Zea mays* ssp. *huehuetenangensis* Doebley is a narrowly distributed teosinte of the western highlands of Guatemala.

Tripsacum: The genus *Tripsacum* is comprised of about 12 species that are mostly native to Mexico and Guatemala but are widely distributed throughout warm regions in the USA and South America, with some species present in Asia and Southeast Asia. All species are perennial, warm season grasses.

a. Species of economic importance to agriculture in Southeast Asia include *Tripsacum dactyloides* and *T. laxum* and *T. dactyloides* (Eastern gamagrass) is a warm season grass native to Mexico and Guatemala but now distributed throughout the Western Hemisphere to Malaysia.

b. *T. andersonii* is native to Central America and it has been suggested to be a hybrid between *Zea luxurians* and *Tripsacum latifolium*

c. *T. lanceolatum* occurs in the Southwest USA and is found along the Sierra Madre Occidental north upto Arizona.

d. *T. floridanum* is native to south Florida and Cuba.

e. *T. manisuroides* is known only from Tuxtla Gutierrez, Chiapas, Mexico.

Various hypothesis have been proposed on the origin/ domestication of maize. Teosintes (*Z. diploperennis* and *Z. mays* sp. *mexicana*) and Tripsacum species are often described as having roles in the domestication process of maize. An early hypothesis proposed that *Z. mays* sp. mexicana was the product of a natural hybridization of Tripsacum and Zea. Further crossings of teosinte with wild maize are thought to have produced the modern races of maize. The possibility of intergeneric hybridization of either *Z. diploperennis* or Tripsacum with extinct wild maize has also been proposed as the ancestral origin of *Z. mays* suggests that domesticated maize may have arisen via human selection of natural hybrids between Tripsacum and perennial teosinte.

Floral biology: Maize is monoecious. The male inflorescence is at the terminal position and is called as Tassel. The female inflorescence is cob or ear which arises in the axil of a leaf.

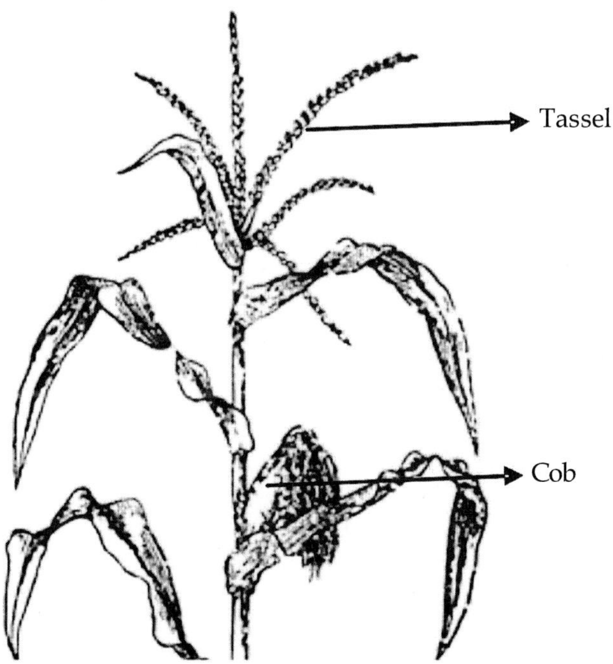

Male Inflorescence (Tassel) is a loose terminal panicle. The central axis of the inflorescence bears paired spikelets in several rows, while the lateral branches bear two rows of paired spikelets. Paired spikelets are similar in structure, one is sessile and the other shortly pedicelled. Each spikelet has two glumes almost equal in size 7-12 nerved, glabrous and bears two florets. Bracteate (Lemma is boat shaped, stiff and parchment like), Bracteolate (Palea is thin and membraneous) perianth is represented by two lodicules, which are fleshy and truncate. Floret is incomplete, unisexual male, cyclic, regular, zygomorphic homo chlamydeous.

Androecium consists of three stamens; filaments are longer than the anthers. Anthers are dithecous, introrse and versatile. In the upper floret, palea is longer than lemma, while in lower floret, lemma is longer than palea.

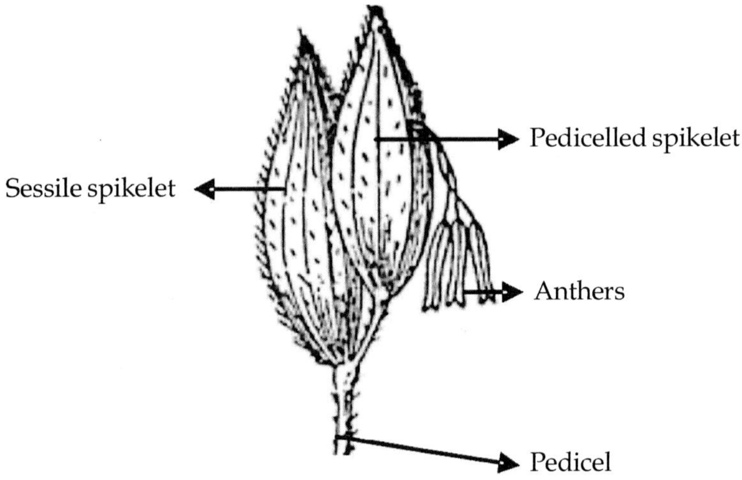

Female inflorescence (cob/ear) is axillary in position. This is a condensed lateral branch, the nodes and internodes are very much condensed and the basal portion is called shank. The inflorescence is covered by a leafy structure callad prophyllium. The other leaves covering the inflorescence are altogether called as husk. The female spikelets are arranged on the fleshy central axis called cob, on which numerous paired spikelets are arranged in longitudinal rows. All the spikelets are sessile and identical. Each spikelet is covered by two glumes which are membranous and consist of two florets. Lower floret is sterile and reduced to lemma and palea which are short broad and membranous with sterile floral rudiments. Upper floret is pistillate, with short broad lemma and palea. Lodicules usually absent, rudimentary stamens may sometimes be found.

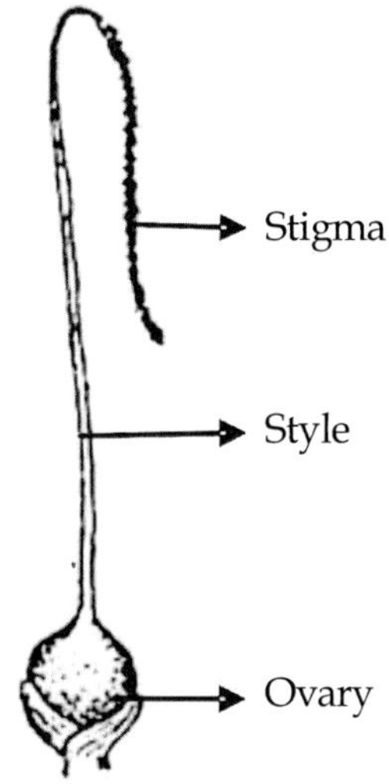

Gynoecium is monocarpellary with superior ovary, it comprises of three carpels. Out of these, two carpels form the style and stigma. The third carpel is the fertile one. Ovary is 'dome shaped', style filiform, very long which grows upto 45 cm in length and emerges from the top of the husks. The silk may be green, yellow, red, brown or purple is solid and flattened or grooved on either side, with short multicellular hairs extending for some distance from the tip, which has two unequal lobes is receptive throughout most of its length. Ovary unilocular with one ovule on basal placentation. Fruit is a caryopsis and seed is moncotyledons, endospermic, endosperm is of various types.

Floral Formula:

Male spikelet: **UF:** Br, Brl, %, P_2, A_3 **LF:** Br, Brl, %, P_2, A_3

Female Spikelet: **UF:** Br, Brl, %, $\underline{G}_{(1)}$ **LF:** Br, Brl, %

Floral Diagram:

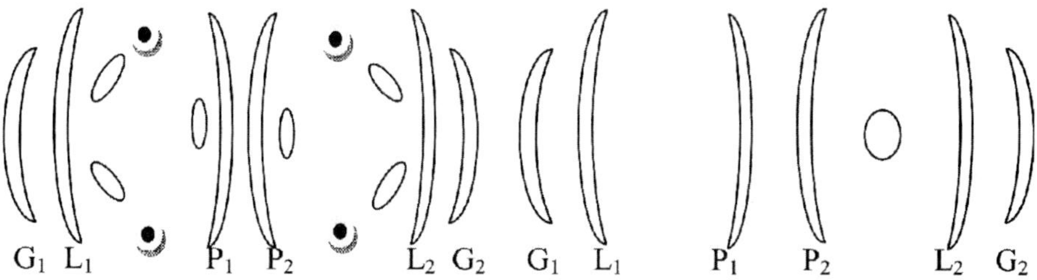

Lower floret Upper floret
Male spikelet

Lower floret Upper floret
Female spikelet

Anthesis and Pollination: Tassel matures earlier than the cob. Flowering starts from the top and proceeds downward both on the main axis and in the lateral branches. The blooming period varies from 4 to 14 days. Usually flowers open between 3AM to 6AM. The emergence of styles takes 2 to 5 days and continues to elongate and remain receptive till it is pollinated. (Style is receptive throughout its length). Selfing is done by covering the female inflorescence with a bag one or two days before the silks emerge out. The tassel of the same plant is also covered with a craft paper bag for collection of pollen and dusting on the emerged styles. For emasculation and crossing, after selecting the desired male and female parents the female inflorescence is covered with a craft paper bag. The tassel of the same female parent is removed and this process is called 'Detasseling'. Pollen from the male parent is collected in the tassel bag and dusted on the silks of female parent, bagged and labeled.

Breeding objectives

1. Yield: The various yield components contributing to yield in maize include number of ears, number of kernel rows, number of kernels per row, test weight and shelling percentage. Grain yield is also dependent on several physiological characteristics such as nutrient uptake, photosynthesis, translocation, sink size and transpiration.
2. Quality: Physical quality of grain is usually measured by kernel hardness, kernel breakage susceptibility (brittleness) and stress cracking. To improve physical quality of maize grain, breeding for grain resisting mechanical damage and drying grain at low temperature is being carried out. Nutritional quality includes protein concentration as well as amino acid balance. The maize protein is called Zein and is low in biological value due to low concentration of the essential amino acids lysine and tryptophan. A breakthrough was made with the discovery of opaque-2 gene (Quality Protein Maize, QPM) which doubled the lysine and tryptophan content in the endosperm and breeding efforts led to the development of opaque-2 based hybrids, synthetics and composites although they have poor agronomic characters which are being improved.
3. Breeding for disease and pest resistance: The most serious diseases of maize are turcium leaf blight, *maydis* leaf blight, post-flowering stalk rot, downy mildew and common rust. Similarly insect pests like stem borers, army worms, aphids, cutworm, jassids, thrips, root worm, leaf miner etc. are important that are causing severe

losses at various stages of crop growth. Understanding the genetic control of these diseases and insect pests as well as the identification of resistance sources against them is important so as to plan an appropriate breeding strategy.

4. Breeding for abiotic stress resistance: Abiotic stresses like water logging, drought and cold stress are appeared as the most critical that affects the maize cultivation.
5. Breeding for adaptability and stability, high yielding baby corn, high oil content and breeding for alternate source of cytoplasm are the other breeding objectives.

Breeding methods

1. Introduction: Many of the introductions have been used directly as commercial cultivars like Sikkim primitive 1, Sikkim primitive 2.
2. Population improvement: Population improvement aims at increasing the frequency of favourable alleles in the population and at the same time maintaining considerable genetic variability. It involves various recurrent selections procedures that are broadly classified into intra-population improvement and inter-population improvement schemes.

   Intra-population improvement: Earlier mass selection was the only method adopted to improve local maize types grown by the farmers. Because of its limitations to identify superior genotypes based on phenotype alone as well no control over pollination and severe inbreeding depression led to the various modifications of the mass selection procedures, which includes, Modified mass selection without pollen control which was also known as Grid method developed by Gardner in 1964. Ear to row selection was developed by Hopkins. In this method, the superior ears are selected from the source population and are kept separate without bulking. In the next year, progeny rows of each individual ear are raised from half the seed and based on superior progeny rows, the remaining seed of the selected progeny rows is bulked to constitute an improved population. Modified ear to row selection proposed by Lonnquist (1964) in which the progenies are evaluated over locations in replicated traits.

   Inter-population improvement: Inter-population improvement involves one/two population and aims at simultaneous improvement of one/two heterozygous and heterogeneous populations. Yield has been the primary trait of selection for the inter-population selection methods and this can be achieved through recurrent selection schemes.

3. Development of synthetic and composite lines: In maize, development of synthetics includes, evaluation of lines on the basis of general combining ability, these selected lines are intercrossed in all possible combinations and equal amount of seed from these crosses is composited to constitute a synthetic and Composite varieties are derived by mixing the seeds of several phenotypically outstanding lines and allowing open pollination among the mixed lines.

4. Heterosis breeding: Instead of using CGMS lines, detasseling the female inbred line is followed because, use of CGMS line is costlier compared to detasseling it is not followed. There are various types of hybrids in maize based on number of inbred lines used which include, single cross, double cross, triple cross, top cross and double top cross hybrids.

5. Transgenic breeding: In maize, transgenic approach would be especially useful for dealing with the problems of stem borer, leaf blight and sensitivity to water logging.

Hybrid maize: Hybrid maize is the first generation progeny from a cross between inbred lines or hybrids among them. In U.S.A, during 1920s and 1930s major efforts in breeding hybrid maize were directed towards development of inbred lines from open-pollinated cultivars and fitting the inbred lines into production of single and double cross hybrid combinations. In 1920, D.F. Jones proposed double-cross hybrid, which became a model for breeding hybrid maize until replaced by the single cross hybrid in the 1960s. in 1950s and 1960s, hybrid seed production was further changed due to the practices like introduction of cytoplasmic male sterility alternative to detasseling practice, and replacement of double cross hybrids with productive single cross hybrids. The male sterile cytoplasm was obtained from open-pollinated cultivar 'Mexican june' and became known as the Texas type cytoplasm or cms-T. Fertility was restored by restorer genes Rf_1 and Rf_2. In 1970, a maize leaf blight disease *Helminthosporium maydis* was virulent on maize hybrids with cms-T cytoplasm. As a result of this experience, the utilization of the cytoplasmic male sterile system to eliminate detasseling in maize was discontinued. Producers of hybrid seed maize then returned to detasseling in hybrid seed production.

List of some of varieties/Hybrids released in India

Name	Parentage	Feature
Introduced maize hybrids		
Sikkim primitive 1		
Sikkim primitive 2		
Mexican line		
Single cross hybrids:		
sangam	Makki safed x tuxpeno PB	_
COH2	UM1810 X UM190	_
Paras	LM5 X LM6	_
DHM 103 (EH 400175)	(CM 120× CM 118) × (CM 119×CM208)	High yielding, yellow, semiflint, bol seeded hybrid tolerant to foliar diseases
DHM 105 (EH 40184)	(CM 119× CM 120) × (CM 113×CM211)	Suitable for all maize growing areas of the country
Rajendra hybrid 2	EVM13 X Jogialocal	_
PEHM 2	CM137 X CM138	Tolerance to water logging
Parkash	CM139 X CM140	Tolerance to lodging
Vivek maize hybrid 4	CM212 X CM141	Tolerance to TLB ,MLB
Shaktiman 2	CML176 X CML186	_
HQPM 7	HKI193-1 X HKI161	_
Vivek QPM-9	VQL1 X VQL2	_
HQPM 5	HKL163 X HKL161	_
	Three way cross hybrids:	
Ganga hybrid makka 4	(CM400 X CM111) X CM602	-
Ganga11	(CM202 X CM111) XCM501	tolerance to DM , TLB
Trishulatha	(CM131 X CM211) X CM120	tolerance to MLB ,TLB
DHM107	(CM119 X CM211) X CM133	tolerance to TLB , MLB
Rajendra hybrid 1	(CM400 X CM300) X P7421	-
DHM109	(CM119 X CM211) X CM133	tolerance to SDM, ESR
HIM 129	(CM128 X CM129) X CM502	-
Mahingha	(KDMI14 X KDMI10) X CM501	-

Shaktiman 1	(CML142 X CML150) X CML186	tolerance to stem borer
Co MH4	(UMI 90 X UMI 285) X UMI112	-
Double cross hybrids:		
VL54	(CM107 X CM 108) X (CM203 X CM204)	-
Ganga-1	(CM101 X CM102) X (CM100 X CM200)	-
Ranjit	(CM103 X CM104) X (CM202 X CM201)	Tolerance to lodging
Deccan	(CM104 X CM105) X (CM202 X CM201)	Tolerance to lodging
Ganga (101)	(CM103 X CM104) X (CM201 X CM105)	Tolerance to lodging
Ganga hybrid makka3	(CM109 x CM110) X CM202 X CM111)	Tolerance to
Deccan 101	(CM202 X CM206) X (CMI15 X CMI14)	Late maturing variety
Ganga 9	(CM202 X CM117) X (CM116 X CM207)	-
Him 128	(CM301 X CM401) X (CM302 X CM303)	-
Deccan 103	(CM120 X CM118) X (CM119 X CM120)	-
Sartaj	(J 617 X H 3191) X (H104-21 X H104-15)	resistant to BLSB
VLMAKKA 42	(U55 B X V 57) X (V20 B X V50)	
Deccan 105	(CM119 X CM 120) X (CM211 X CM107	Tolerant to MLB TLB , WILT
CoH3	(UMI101 X UMI130) X (UMI90 X UMI285)	Resistant to downy mild dew , tolerant to stem borer
Composites:		
Renuka	G15 X G1Y	-
Amber composite	Composed of eleven flint × dent varietal crosses	Long duration, broad dark green leaves, goon fodder.
MCU 508	Indian pools	-
Arun	Indigenous & exotic material	-
Parbhat	-	-

Harsha	Intermix of various crosses	Tolerance to TLB
PUSA-1	Pool of pop A600, A602, A603, A604	-
PUSA-2	Pool of A61	-
VL MAKKA 88	National pool of CD yellow	-
Kiran	-	Tolerance to BLB
Varun	Compositing of ten inbreeds	-
Azad uttam	Compositing of 250 open pollinated cases of various trials of KH 1997	Early maturing variety
Panchganga	Pool 15	Suitable for mixed cropping
C-14	-	Tolerance to lodging
C 8	It is pool of A3, Jellicrose, thana, mandi-598	Resistance to lodging, stem borer
Birsa makal	D741 & gene pool of early yellow	Resistance to MLB
Mahl dhawal	Broad based gene pool of HSN (RFC) K-87	Bold and moderate resistance to DM
Dewaki composite makka	Pool of American dent, tuxpeno	Resistance to fusarium wilt
Jawahar makka 8	Early to very early national pool material	Resistance to MLB
Jawahar composite-12	Early to very early national pool	Tolerant to TLB , MLB
Gaurav	OPV selected from various trials	Resistance to major foliar and stem diseases
Amber popcorn	Synthetic	Suitable for popping

BLSB –Banded leaf sheath blight

TLB-turcicum leaf blight

DM-downy milddew

ESR-erwinia stalk rot

MLB-maydis leaf blight

Economic importance: The grain is used as food for man, and livestock. The immature cobs are largely eaten after roasting or boiling. The grain is also used in making corn starch and industrial alcohol. Corn oil is used for soap making, lubrication and as a salad oil.

❑❑❑

Chapter-6

Breeding in Millets: Sorghum, Pearl Millet and Finger Millet

Sorghum (2n = 20)

Sorghum (*Sorghum bicolor*) is a major crop ranks fifth after rice, wheat, maize and barley and is an important component in traditional farming systems and in diets of millions of people. It is one of the most important food crop in the semi and tropics of India, Africa. Australia, Argentina, mexico. It belongs to the genus *Sorghum* and the tribe Andropogonaea of the family Gramineae (Poaceae). Center of origin of sorghum is Ethopia and South Asia. The progenitor of *S. bicolar* is *Sorghum arundinaceum.*

Wild relatives of the genus Sorghum: Snowden (1935 and 1936) and Garber (1950) suggested that the genus Sorghum comprises six subgenera (later named "sections"):

1. *Eu-Sorghum* – It is the likely progenitor of *S. bicolor* (n=10) in Africa, of the n=20 chromosome *S. halepense* in India and of *S. propinquum* (n=10) in South East Asia. It has been suggested that crosses between *S. propinquum* and *S. bicolor*, when the latter reached China, gave rise to the distinct Chinese landraces. *S. propinquum* is rhizomatous so is considered to be a progenitor of the strongly rhizomatous and geographically neighbouring, *S. halepense.*
2. *Sorghastrum* (Nash) – *Sorghum* and *Sorghastrum* probably had common ancestors. *Sorghastrums* were found in Africa and the Americas
3. *Chaetosorghum* – found only in Australia
4. *Stiposorghum*- found only in Australia
5. *Heterosorghum* - found in South East Asia, Phillipines and Australia
6. *Para-sorghum* - widely spread in South and East Africa, India, South East Asia, Australia, and as *S. trichocladum* in western Mexico and Guatemala

This classification later refined by Garber (1950) and Dogget (1970). They classified Sorghum in to two sections, Section I Sorghum (True sorghum) and Section II Para sorghum. True sorghum further classified into two sub sections Arundinaceae and Halepensia. Sub section Arundinaceae again classified into two series Spontanea (grass) and sativa (grain). Series spontanea consists of the species *S. sudanense, S. aethiopicum, S. virgatum, S. verticillifoilum, S. stapfii.* Series sativa consists of the species *S. vulgare, S. subglabaesence, S.dochna.* Subsection Halepensia consists of *S. halepense, S. miliaceum, S. almum, S. propinquum, S. radolphianum.* And section II para sorghum consists of the pecies *S. versicolor, S. introns, S. purpureosericeum, S. nitidum and S. plumosum.*

Later De Wet (1978) considered the *Sorghum* subgenus comprise three species, *Sorghum halepense, S. propinquum* and *Sorghum bicolor.* The cultivated taxa of the subgenus *Sorghum* were first grouped into 28 species by Snowden (1936). Classification schemes since then have all been based on his historic work. Later all named *S. bicolor* (L.) were grouped into working groups (WG) by Murty and Govil (1967) and later into five races and ten intermediate groups by Harlan and de Wet (1972). This is mainly based on the coverage of the glume in the grain.

1. Bicolor (B): Grain elongate, glumes clasping the grain which may be completely covered or ¼ exposed.
2. Guinea (G): Grains flattened dorso-ventrally.
3. Caudatum(C): Grains asymmetrical, glumes 1/2 the length of the grain.
4. Kaffir (K): Grains symmetrical (spherical), glumes clasping in varying length.
5. Durra (D): Grains rounded obovate, wedge shaped at the base and broadest slightly above the middle; glumes very wide.

Recently Dahlberg (2000) integrated the WG and race/ intermediates grouping system for use by researchers who require a more detailed description of groups. Most sorghum workers recognize and use the race/intermediate method of grouping. Historically, the Kafir, Caudatum and Durra races have contributed most to grain breeding. For example, the cytoplasmic-genetic male-sterility system almost exclusively used to produce F_1 hybrid cultivars is based on the Milo cytoplasm (A_1) from the Durra race and to the non-restorer genes from Kafir. A_1 cytoplasm male-fertility restoration genes are ubiquitous across the races.

Floral biology: Inflorescence is Terminal panicle or head either compact, or loose (lax). Axis angular, hairy, Peduncle is called 'Neck'. When it is curved down it is called 'Goose-necked'.

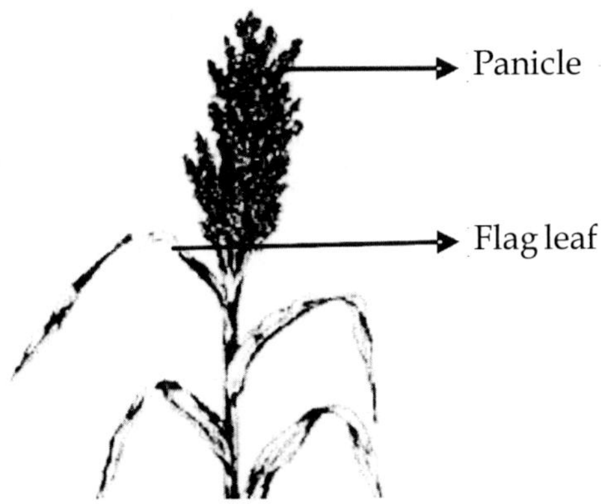

Spikelets are borne on ultimate branches, generally on tertiary branches. Spikelets occur in pairs or in three numbers at the end of the branches. In paired spikelets one is sessile and the other one is pedicellate. In case of three spikelets one is sessile and the other two are pedicellate.

Sessile spikelet: It is fertile and only sets grain. It has two glumes and two florets. The lower floret is sterile, consisting of only lemma in the form of a broad ciliate membranous bract which partially enfolds upper perfect floret. The upper floret consists of a membranous lemma, with a short or long arm, a small thin delicate palea. This may be absent. Perianth is represented by two lodicules present adjacent to fertile lemma, short, broad, fleshy, truncate, with ciliate margins.

Pedicellate spikelet: It is smaller in size than the sessile. Spikelet often sterile rarely staminate. It has 2 glumes and 1 or 2 lemma. When there are two lemmas lemma (L_1) is epaleate, elodiculate and sterile, lemma -2 (L_2) is paleate, lodicuiate and staminate.

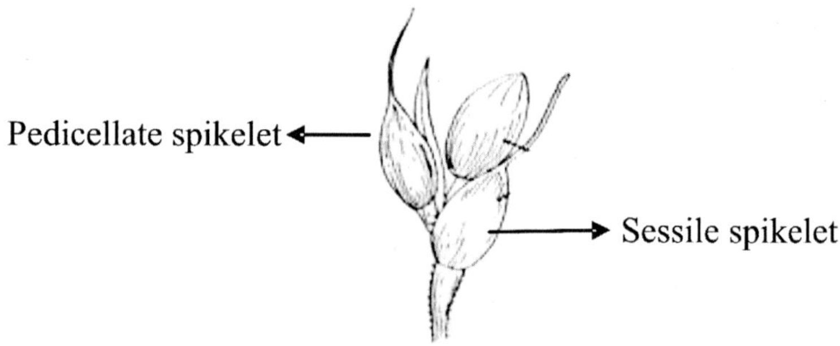

Androecium consists of three stamens, sagittate with long anthers, dithecous, introrse and versatile anthers. Gynoeeium is monocarpellary, unilocular, superior ovary with a single ovule on basal placentation styles two terminate into plumose stigma. Fruit is a caryopsis, wholly enclosed between glumes.

Floral Formula:

Sessile Spikelet: Br, Brl, %, ⚥, P_2, A_3, $\underline{G}_{(1)}$ (Upper floret) and

Br, Brl,– – – – –(Lower floret)

Pedicellate spikelet: Br, Brl, %, , P_2, A_3, (Upper floret) and

Br, Brl,– – – – –(Lower floret)

Floral Diagram:

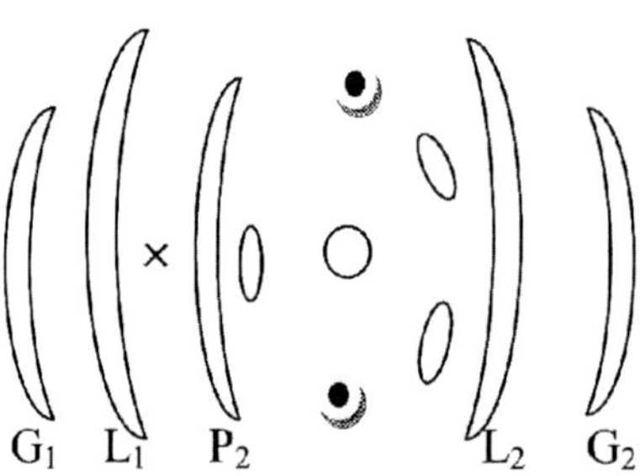

Sessile spikelet

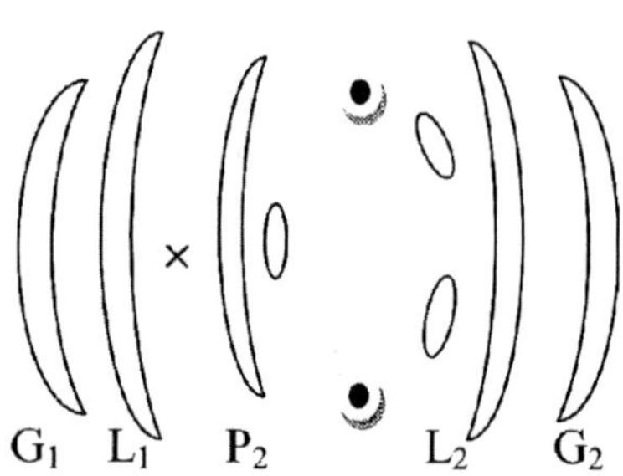

Pedicellate spikelet

Anthesis and Pollination: A panicle takes about 2 to 3 days to completely emerge from the boot leaf. Flowering starts from tip of the main axis and proceeds downwards. A panicle completes flowering in about 8 days. Sessile spikelets are first to open followed by pedicellate ones. Blooming occurs from mid-night to early morning. Anthers and stigmas come out simultaneously. Anthers desiccate as soon as they come out and shed pollen on the stigma. The pollen loses its viability within a few hours (A single panicle may releases 25 to 100 million pollen grains). The stigmas are receptive for one or two days before the flower blooms and for several days after blooming. Extent of cross pollination range 5-50%. Selfing is done by covering the just emerging panicle with craft paper bags. Emasculation can be done using forced open method or hot water treatment. Another method is pencil technique. In this tip of pencil is placed at the base of spikelet and rotated pointing upward, due to the pressure created the anthers come out from the top of the spikelet and remove the anthers using forceps. For crossing, pollen is collected in bags and dusted over the exposed stigmas or a pollen producing head may be brushed over an emasculated head. Emasculated heads or heads that have been pollinated are enclosed in bags to protect them from stray pollen.

Breeding objectives

1. Yield: Achievement of both high grain and forage yield mainly depends on components like Panicle length and breadth panicle weight, number of primary branches, number of seeds/panicle and 100 seed weight, plant height, leaf area index endosperm texture.
2. Quality: Sorghum grain is used for food and livestock feed, sorghum plant is utilized for forage. This diversity of uses increase the complexity of breeding of quality traits like bread making, redgrain for biscuit making, protein and lysine content and low HCN content are important.
3. Resistance to lodging: Lodging resistance was improved by utilization of dwarfing genes to breed for short stature, by crossing to kafir and hegari to obtain stout stalk.
4. Early maturity: Breeding early maturity permits sorghum escapes from drought and other stresses.
5. Resistance to biotic stresses: The most serious diseases of sorghum are Grain mold, charcoal rot, anthracnose, downy mildew, rust and smuts etc. In breeding for disease resistance, a screening procedure is needed that includes inoculation of sorghum genotypes with causal pathogen and evaluation of disease infection either in the field or green house. Like that serious pests are stem borer, midge and green bug etc and also for getting insect resistance the main target is nonprefrence or reduced reproductive capability of insect are known as antibiosis.

6. Resistance to abiotic stresses: Heat and drought stress are important criteria in sorghum. Mechanisms that contribute stress tolerance include extensive root development to maximize water supply, smaller and fewer stomata to reduce leaf water loss etc.
7. Other breeding objectives include sweet sorghum, striga resistance and isolation of alternate source of cytoplasmic genetic male sterile source.

Breeding methods

1. Introduction: Varieties of milo and kafir sorghum introduced from USA are used in conversion programme to convert the local long duration photo sensitive varieties to short duration, non-photo sensitive lines.
2. Selection: By pure line selection, improved strain shows superiority over the existing variety, it is released for commercial cultivation on large scale. E.g M-35-1, Yashda Maulee (RSLG-262).
3. Hybridization followed by selection which includes inter varietal and inter specific hybridization. Pedigree method was followed in the development of varieties SPV-86 (R-24 X R-16), and CSV -15 R, (SPV-475 X SPV-462).
4. Back cross method: It is used for conversion of tall, late maturing, tropical accessions from world sorghum collections to short, day-neutral genotypes adapted to temperate climates. Tropical cultivars are converted to temperate climate adaptation by substituting two recessive alleles for height for the dominant counterpart alleles in the tropical accession and the recessive ma_1 maturity alleles for dominant Ma_1 through a back cross programme.
5. Population improvement: Population improvement involves the use of recurrent selection to increase the frequency of genes for a particular quantitatively inherited character. The various procedures followed are mass selection, half sib family selection, full sib family selection and S_1 family selection etc.
6. Transgenic breeding: Transgenic research is mainly needed to develop lines with resistance to grain mould, stem borer and shoot fly etc.
7. Other breeding approaches like mutation breeding and use of apomictic lines for crop improvement in sorghum.

Hybrid sorghum: With the discovery of cytoplasmic male sterility and fertility restoring genes in sorghum in the early 1950s. The tools needed for commercial production of hybrid sorghum became available. Cytoplasmic male sterility was obtained by backcrossing chromosomes of kafir into the cytoplasm of milo. Fertility was restored in the hybrid when the CMS plants were pollinated from a milo plant. Cultivars of kafir origin then in commercial production were readily converted to CMS, and cultivars of milo origin were used as pollinators. In the commercial production of hybrid seed, the A×R hybrid is produced by planting alternative rows of A and R lines in an isolated field. Twelve rows of the male sterile A-lines are commonly planted to four rows of the pollinator R-line.

List of some varieties/hybrids released in India

Varieties	Parentage	Special characters
CSH-1	CK 60 A×IS 84	-
CSH-5	2077 A×CS 3541	Rust resistant, foliar disease resistant.
CSH-6	2219 A×CS 3541	-
CSH-9	296 A×CS 3541	-
ICSV-1	SC 108-3 × CS 3541	-
NTJ 1 (NJ 2092)	148 × 296 B	Suitable for early Rabi tract, tolerant to drought and non lodging
N 14 (NJ 1031)	Pure line selection from proddatur local	Yellow jowar with large compact ear heads
NTJ (NJ 2122)	Pure line selection from zera zera germ plasm of ICRISAT	Early and drought tolerant. Bol pearly white grain.
Pusa chari 9	-	-
Rajasthan chari -1	-	-
Pusa chari-23	-	-
PSH-106	-	-
MP Chari	-	-
Rajasthan Chari-2 (SU-45)	-	-
CO-27	-	-
HC- 260	-	-
HC-171	-	-
AS-16	-	-

CSH-13K8R	296 A×RS 29	-
CO H4	296 A×TNS 30	Fewer incidences of leaf diseases, grain mould and sugary disease.
CSH-14	AKMS 14A×AKR 150	-
UPFS-23	-	-
SSG -988	-	-
CSV-15	-	-
GFSH-1	-	-
MFSH-3	-	-
L×250	-	-
Harasona 855F	-	-
HC-308	-	-
CSH-16	27 A×C 43	-
M 35-1(C)	R 4×R 16	Resistance to shoot fly, incidence of charcoal rot.
FSH-92079	-	-
CO- S- 28	CO-25×SPV-942	Resistant to shootfly,ear head bud and grain mould
CO-5	ICS-51 A×TNS-30	Tolerant to shoot fly, grain mould
NJ 2401	NTJL×CMS-3	Tolerant to lodging,good texture, baking quality
Johar	-	Suited to rain-fed area
CO-26	MS 8271×IS 3691	-
CO-30	APK-1×TNS 291	Resistant to downy mildew,have creamy white grains
SPH-1648	MDSH263×CSH-29	Resistant to shoot fly and charcoal rot
Paiyur-II	Selection from IS 15845	Suited to rain-fed tract,tolerant to charcoal rot
Paiyur-I	CO-19×CO-24	Photo sensitive
K-Tall	2219-A×IS-3541	-
K-11	K-7×A-6552	Tolerant to drought, non- shattering
APK-1	Hybrid derivatives of TNS-30×CO-26	Non-lodging
CSV-1	Selection from IS-3924	White mid rib, non -sweet
CSV-2	IS 3922×Karod local	-
CSV-3	IS 2954×BP-53	-
CSV-4	IS 3675×IS 3541	Resistant to leaf disease,leaf mould,downy mildew

CSV-5	IS-398×Aispuri local	Resistant to head mould , downymildew
CSV-6	IS 3922×Aispuri local	-
CSV-7R	IS 2950×M 35-1	-
CSV-8 R	R-16×R 24	-
CSV-9	CS 3541-Tall mutant	-
CSV-10	SB-1066×CS 3541	-
CSV-11	SC 108-3×CS 3541	-
CSV-12	(2947×232)× (I S3612×2219B)	-
CSV-13	(ISI 12622×555)×(IS 3612× 2219 B)×(E-35-1-52)	-
Swati	SPV 86 × M 35-1	-
PSH 1	PSA 3 × PSR 23	Non lodging medium duration hybrid suitable for cultivation in red and medium heavy soil.
PSV 1 (SPV 462)	MS 827 × IS 3691	Non lodging and moderately tolerated to grain moulds. Good keeping quality and roti making quality
ASH 1	PSA 9 × PSR 34	Tall, non lodging, dual purpose, long duration variety
Palem sorghum 2 (SPV 1293)	SPV 86 × GD 57904	Dual purpose, non lodging, medium bold and pearly white grain. Tolerant molds and leaf diseases.
CSH 13-R	-	-
SAMSORG-8 (ksv-14)	L 9289×C.E 90	Photo-insensitive,open panicle resistant to major leaf disease.
SAMSORG-38 (NR-7-1176)	Introduced from sudan through ICRISAT	Semi dwarf compact panicle, striga resistance

Economic importance: Used as a food (millet) crop in the drier parts of tropical Africa, India and China. The threshed grain is ground into a who i.e. meal flour and used for making thin porridge.

Pearl Millet (2n = 20)

Pearl millet (*Pennisetum glaucum*) is the fifth most important cereal crop in the world after rice, wheat, maize, and sorghum. It is a widely grown rainfed cereal crop in the arid and semi-arid regions of Africa and Southern Asia. In sub-Saharan Africa, pearl millet is the third major crop with the major producing countries being Nigeria, Niger, Burkina Faso, Chad, Mali, Mauritania and Senegal in the West and Sudan and Uganda in the East. It belongs to the genus *Pennisetum* and the tribe *Paniceae* of the family Gramineae (Poaceae). Center of origin of pearl millet is Western Africa. The progenitor of *Pennisetum glaucum* is *Pennisetum purpureum*.

Wild species: The genus *Pennisetum* appears to have more than 100 species, divided in to five sections: Gymnothrix, Brevivalvula, Heterostachya, Eupennisetum, Pencillaria. Pearl millet and its closest wild relatives are native to west and east Africa and are now part of section Pencillaria. Brunken *et al.,* (1977) scored diverse genepools for a number of floral and grain characteristics and identified grain shape as the most consistent trait, which follows a geographic pattern. Four basic grain shapes were found in the world collection and were used for classification, although not all pearl millet accessions fit neatly in to any of these four basic races.

Race typhoides: The race typhoides is characterized by obovate caryopsis that are obtuse and terete in the cross-section. Inflorescences are mostly cylindrical in shape. Morphologically it is the most variable among the four races and is also most widely distributed. It occurs across the entire African continent. It is the only basic race found outside Africa and the predominant race grown in India.

Race nigritarum: In this race, caryopsis is angular in cross-section with three and six facets per grain. Inflorescences are candle-like. The apex of the grain is usually truncate and often tinged purple. The mature grain is generally longer and protrudes beyond the floral bracts. This race is generally found in western Sudan to northern Nigeria.

Race globosum: The caryopsis in this race is spherical with each of its dimensions being approximately equal. Depth of the grain always exceeds 2.4 mm. The grain is otherwise terete and obtuse. Inflorescences are candle shaped and often exceed 1 m in length. It is the most common race in Benin, Ghana, Niger, Central Nigeria and Togo.

Race leonis: The race leonis is characterized by an acute, oblanceolate, terete caryopsis. The most distinct character of the leonis grain is its acute apex, which is terminated by the remnants of the stylar base. At maturity, approximately one-third of the grain protrudes beyond the floral bracts. Inflorescence shape is candle-like. It is specific to Sierra Leone but also grows in Senegal and Mauritania.

Thus the four cultivated forms of pearl millet i.e *typhoides* found mainly in India and Africa, *nigritarum* dominant in eastern Sahel, *globosum* dominant in the western Sahel and *leonis* dominant on the West African coast. Recently different taxa of the pearl millet were classified according to their closeness to the cropped species *P. glacum* is as follows, Primary wild relatives (*P. glacum* cultivars and landraces, *P. glacum* subsp.monodii and *P. glacum* subsp. Stenostachyum, *P. violaceum* and weedy form *P. sieberanum*), Secondary wild relatives (*P. pupureum* and *P. squamulatum*) and Tertiary wild relatives (all other species in the genus).

Floral biology: Inflorescence is terminal contracted panicle, cylindrical spike. Spikelets borne at the tip of short lateral axis which are covered with white silky hair. Spikelets occur in group of one to three subtended by 10-50 scabrid bristles, which are longer than spikelet. Spikelets are small and lanceolate. Each spikelet has two glumes, which are obscure and two florets. Lemma-1 is paleate, elodiculate and staminate with penicellate anthers. Lemma-2 is paleate, lodiculate and bisexual.

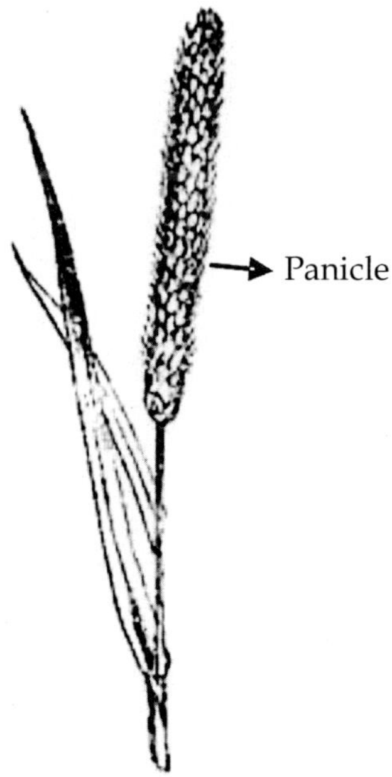

Androecium consists of three stamens, penicellate, anthers are dithecous, introrse and versatile. Gynoecium is monocarpellary, unilocular superior ovary with one ovule on basal placentation. Styles are two connate at the base with plumose stigma. Fruit is a caryopsis.

Floral formula:

Upper floret: Br, Brl, %, ⚥, A_3, $G_{(1)}$

Lower floret: Br, Brl, %, ⚥, P_2, A_3

Floral diagram:

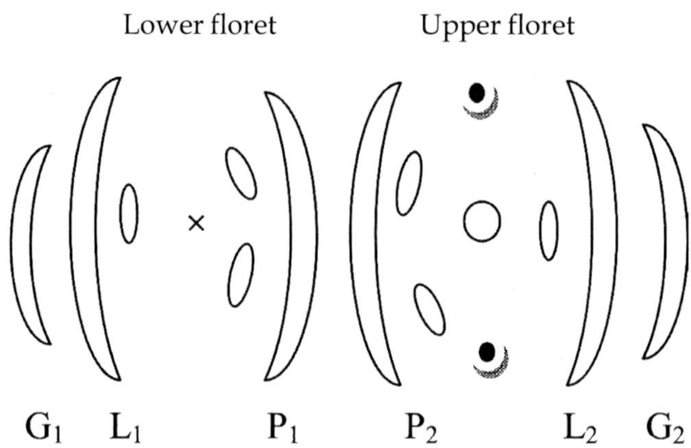

Anthesis and Pollination: It is highly cross-pollinated crop due to protogynous condition. The style begins to appear 2 to 3 days after emergence of spike. First at 1/3 rd portion and gradually downward, and style remain receptive for 1 to 2 days. Anther emerges after styles dry up. The anthers of bisexual flower appear 2 to 3 day before those of staminate flowers. The maximum emergence of style is between 9.00 A.M to 3.00 P.M. while maximum opening of the flower occur between 10.00 A.M. to 2.00 P.M. and minimum between 3.00 and 6.00 P.M. To ensure selfing spike may be bagged before emergence of the stigmas. Hand emasculation is laborious and difficult due to small size of flowers and the late development of anthers in relation to stigma. Protogynous nature of the flower suggests the easy method of crossing without emasculation. In the lowermost part of the ear head the styles appear very later. In the lowermost part of the ear head the styles appear very late. This interval between appearance of styles and dehiscence of anthers is more particularly in lowermost region of the spike. Hence the spike emerged from the flag leaf is selected for emasculation and upper 2/3 rd or 4/5 portion of ear head is cut and remaining portion of the head is bagged immediately. It is always done in the morning after 2 to 3 days when styles appear, fresh pollen grains collected from protected male parent in petridish and dusted over the stigma of the bagged heads and rebagged immediately. The pollination may be repeated 2-3 times till the anthers of the same region show their appearance.

Breeding objectives

1. Yield: Achievement of high yield mainly depends on components like more number of tillers, well filled compact long panicle, heavy grains and uniformity of ripening.
2. Quality: It can be achieved by incorporating yellow endosperm to improve vitamin A content or white endosperm to improve protein content.
3. Resistance to biotic stresses: Downy mildew, ergot, rust, blast, mycotoxins and smut are major diseases. Lines having Local Bellary cytoplasm (732A) are observed to be downy mildew resistant.
4. Resistance to abiotic stresses: Drought is the major problem. This can be achieved through developing lines having shorter duration so that they can escape from drought and also lines with high leaf water potential and more adventitious roots and high chlorophyll stability index are need to be developed.
5. Breeding for high forge value: Forage must have the traits like high sugar content in the stem juice, increased leaf number with more breadth and digestability.
6. Other breeding objectives includes alternative source of cytoplasm because of highly susceptibility of Tift cytoplasm to downy mildew, bold grains and ripening uniformity, reduced plant height and bristeled ear etc.

Breeding methods

1. Introduction: Male sterile cytoplasm Tift 23 A introduced from USA.
2. Selection: Mass selection is commonly followed. The criteria for selection are usually head characteristics, like compactness, length of ear, weight of grain, uniformity of ripening etc. Varieties developed are Jamnagar, Jaint, Pusa moti, etc.
3. Hybridization and selection: Interspecific hybrids between napier grass and bajara are Napier bajara hybrids, the popular hybrids are

Desharat and Yashawant famous for high yield. These are also known as elephant grass.

4. Population improvement/synthetics/composites: Population improvement mainly emphasized to avoid problems related with narrowing down of the genetic base of the crop. The synthetics are reconstituted at certain intervals by inter pollination among the component lines. E.g. ICMS-7703, ICMV-87901.
5. Mutation breeding: At IARI Tift 23 A was gamma irradiated and 5071 A resistant to downy mildew was evolved.
6. Transgenic breeding: It is mainly used to get transgenics for resistance to downy mildew.

Hybrid bajra: In earlier days before the identification of male sterile lines utilising the protogynous nature hybrids were released. The hybrids were produced by sowing both parents in the ratio of 1:1. After the discovery of cytoplasmic genic male sterile line Tift 23A by Burton in Tifton, Georgia led to development of hybrids. Earlier hybrids of India *viz.,* HB1, HB2 to HB5 were produced utilising Tift 23 A. But due to susceptibility to downy mildew they went out of cultivation. Even before the discovery of CGMS lines by Burton it was discovered by Madhava Menon and his coworkers at Coimbatore. Unfortunately due to failure of publishing it was not recognised. To over come the problem of downy mildew male sterile lines L 111A and 732 A were isolated and at present used in breeding programme.

List of some varieties/hybrids released in India

Variety	Parentage	Special traits
Pusa Composite 612 (MP 480)	-	Medium maturity, medium to tall plant height, compact cylindrical earheads
AIMP 92901 (Samrudhi)	-	80 days, medium height, cylindrical compact earheads slightly tapering towards tip, bold deep grey grains
JBV-2 S.	-	79 days, tall, compact cylindrical earheads,yellow anthers, obovate brown grains, developed from Early Composite 91 (EC 91)
Pusa Bajri 266	-	80 days, tall, thick, semi compact, cylindrical ear head, yellow anthers, obovate grey brown

Anantas (APS-I)	-	85 days, medium tall, long cylindrical earheads, light grey grains
ICMV 221	-	75-80 days, medium tall, thick lanceolate semicompact earheads, purple anthers, globular dark grey grains, bred from C3 cycle of the bold seeded early composite (BSEC)
Nagarjuna	-	75-80 days, tall, good tillering, semi-compact earheads, developed from 20 elite lines of African and Indian origin
Balaji	-	75-80 days, medium tall, dark green foliage, long cylindrical earheads, bold light grey grains
New vijay	-	80-85 days, tall, good tillering, developed by half-sib and mass selection
Vijay	-	85-90 days, tall, pigmented nodes, long compact cylindrical earheads, bold grains, a composite variety from Senegal selection
Pusa safed	-	83-87 days, medium tall, semi-compact earheads, yellow anther, yellowish white grains of medium size
Pusa composite 334	-	75-80 days, tall, thick ,semi compact cylindrical earheads, obovate grey brown grains, developed from three selected lines and eight elite inbreds
PPC-6 PARBHANI SAMPADA)	-	75-80 days, medium tall, light pink nodes, bold grains, bred from inbred lines
Hybrids		
86M89 (MH1747)	M163F X M164R	Late maturing, tall height, conical very compact earheads, yellow anther colour, resistant to downy mildew, obovate deep grey coloured Seed
HB 4	TIFT 23A × K 560	85-90 days medium tall, profuse tillering, rod - shape earheads, greenish yellow grains
NHB 5	5071A × K 559-85	85 days, medium tall, thick stems, cylindrical earheads, medium bold light

PHB 14 (HB 7)	PB 111A × PIB 228	87 days, tall, thick stem, long bristles ear heads, broad leaves
NHB-3	5071A × J104	75-78 days, tall, thin stem, 4-6 tillers, narrow hairy leaves, spindle shaped ear heads
HB 1	TIFT 23A × BIL 3B	80-85 days, medium tall, profuse tillering, succulent stem, green ligules, long earheads,
BJ 104	5141A × J104	75-82 days, medium tall, good basal tillering, nodal joints red, medium long cylindrical earheads tapering towards tips, conical grains
BD 111	5141A × D 111	85-90 days, tall, good tillering, small bristles, resistant to lodging
KBH 1	PB 111A × S 444	85-90 days, tall, white mid rib, short bristles, grey grains
PUSA 763 (BD 763)	5141A × D 763	90-95 days, tall, thick stem, broad long leaves, rod shape earheads
MBH 110	MS 2 × POLLINATOR NO. 2	85-87 days, medium tall, loose and thick conical earheads with long bristles, bold grey shining grains
PUSA 46 (CM46)	5054 A × M 46	75-78 days, tall, thin stem, 4-6 tillers, narrow hairy leaves, spindle shaped earheads
X 4	5141A × PT 1921	80-85 days, tall, purple pigmentation at the base of stem,
GBH 27	5141A × J 2002	80-85 days, medium tall, nodal pigmentation, long and cylindrical earheads, bold grains
PHB 47	PB111A × PIB 1234	85 days, tall, thick sturdy stem, long thickear heads with profuse gristles, remains green till harvest
HHB 60	81A × H77/833 – 2	74-76 days, tall, high tillering, resistant to drought and salt stress and good quality fodder
MHB 136	MS 2 A × PL NO. 6	85 days, medium tall, high tillering, compact conical vary thick ear heads, yellow anther, pink medium long bristles, grey seeds
VBH 4		VBMS 1A × VBR 19 80 days, medium tall, good tillering, compact non–bristled earheads, medium bold

ICMH 312	81A × ICMR 312	85 days, medium tall, long semicompact conical earheads, yellow to purple anthers, globular grey grains
ICMH 356	ICMA88004 × CMR 356	75-80 days, medium tall, semi-compact anthers, thick conicalear heads, yellow obovate bold yellow brown grains
PROAGRO 1 (FMH 3)	PSP 21 X PP 23	85-88 days, tall, excellent tillering,
GK 1004	GKPM 1A X GKPM 59R	85 days, medium tall, semi compact medium long earheads, grey grains
APBN 1 (Forage bajra)	IPM 12159 × Kasate 52440	High yielding with yield potential of 300 tons green fodder per/ha/annum and 60 tons dry fodder per/ha/annum
Mukta (PBS 2)	Synthetic	Suitable for rain fed cultivation, tolerant to drought.
Mallikarjuna (PBS 1)	Synthetic	Tolerant to downy mildew, ergot and small diseases

Economic importance: Used as staple food in the drier parts of tropical Africa and in India. The green plants produce a useful fodder and it is sometimes grown for this purpose.

Finger Millet (2n = 36)

Finger millet (*Elusine coracana*), among the millets it ranks fourth in importance after sorghum, pearl millet and foxtail millet. It is widely cultivated in north and central Europe, Africa, temperate Asia (western Asia, Arabia and China) and tropical Asia (India, Indo-China and Malesia), Australia and the southwestern and northwestern Pacific. It belongs to the genus *Elusine* of the family Gramineae (Poaceae). Center of origin of finger millet is Africa (according to Vavilov) and India (according to Decandole). The progenitor of *Elusine coracana* is *Elusine ntermod*.

Wild relatives: The genus *Elusine* comprises of 11 species of which six are diploids and five are tetraploids. It is now evident that cultivated finger millet is highly variable within its primary centre of origin in Africa and secondary centre in Indian sub-continent. Both wild finger millet (*Eleusine coracana* subspecies *ntermod*) and the cultivated finger millet (*Eleusine coracana* subspecies *coracana*) are important from the point of view of germplasm collection and conservation and form the primary genepool. Wild finger millet (subspecies *ntermod*) is native to Africa but has migrated to several warmer parts of Asia and America. The diploid wild species *E. indica, E. floccifolia* and *E. tristachya* form the secondary genepool and the tertiary genepool comprises species *E. intermedia, E. gaegeri, E. kigeziensis, E. multiflora* and *E. semisterlis (E. compressa)*. The cultivated finger millets have been divided into four races mainly based on the inflorescence morphology.

Race Elongate: Long slender inflorescence, branches 10-24 cm long, digitately arranged spikes spreading and curved outward at maturity. In this three sub races *laxa, ntermo* and *sparsa* are recognized.

Race Plana: Arranged in two more or less even rows along the rachis giving the inflorescence branch a flat ribbon like appearance. In this three sub races *seriata, confundere* and *grandigluma* are recognized

Race Compacta: Spikelets are composed of 9 or more florets and incurved at the tip to form a large fist-like inflorescence.

Race Vulgaris: Most common type with 4-8 florets in a spikelet giving an appearance of semi-compactness with tip of finger incurved at maturity. In this four sub races *liliaceae, stellata, incurvata* and *digitata* are recognized.

Floral biology: Inflorescence consists of terminal digitate spikes of spikelet, borne on a long peduncle from the end of which four to five spikes radiate in a whorl called fingers with an odd one a little lower down the whorl called as thumb. Rachis of the spikes are flat, spikelets are sessile, arranged in two rows alternately attached to one side of the rachis. Each spikelet is having 3-7 florets enclosed by common glumes, dorsally compressed, each spikelet consists of an outer glume which is small and the inner glume as long as the spikelet and 4-6 florets. The top most floret is either sterile or staminate. The florets are present on zig-zag rachilla. Lower florets bigger in size with lemma boat shaped with green median nerve, margin over laps the edge of palea. Palea keeled with two nerved pointed, lodicules two, small truncate. An average number of spikelets on a finger had been found to be 67 to 73.

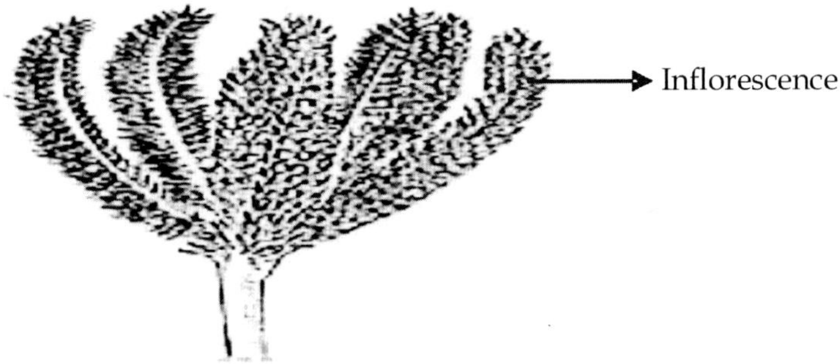

Androecium consists of three stamens with dithecous, introrse and versatile anthers. Gynoecium is monocarpellary, unilocular, superior ovary, with one ovule on basal placentation. Fruit is utricle.

Floral formula:

Upper floret: Br, Brl, %, ⚥, P_2, A_3, $G_{(1)}$

Lower floret: Br, Brl, %, ⚥, P_2, A_3, $G_{(1)}$

Floral digram:

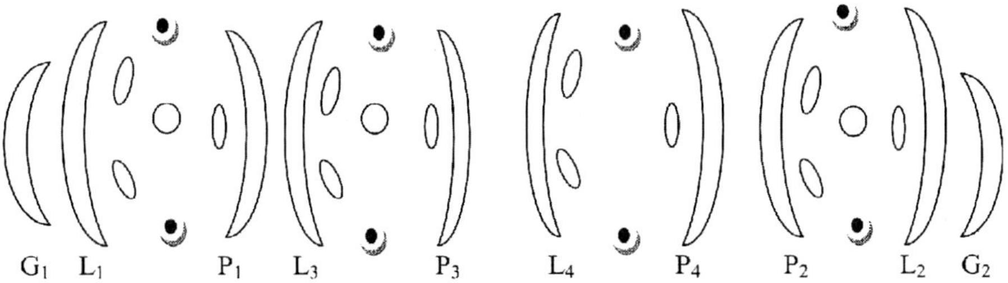

Anthesis and pollination: In finger millet, complete emergence of inflorescence requires 7-8 days and anthers require 45 minutes for dehiscence after emergence. Stigma is receptive for about five minutes after emergence from the glumes. And also depending upon the ear shape, flower opening period will vary i.e. for compact 2-3 A.M, fisty 3-5 A.M and open 1-2 A.M. In this crop, pollen viability is very short for 10-15 minutes. Self pollination is the rule because the period of anthesis is very short but cross pollination by wind and insects is less than one percent. In ragi, commonly two methods of emasculation are using, hand emasculation and hot water treatment. Hot water treatment at 52°C for 2 minutes was the best as judged from the percentage of seed set.

Breeding objectives

1. Yield: Yield is important target in every crop.
2. Quality: Breeding for high protein white ragi varieties suitable for malt making
3. Early maturity: Breeding for short duration varieties suitable for rainfed conditions is important objective
4. Resistance to biotic stresses:
5. Resistance to abiotic stresses: Breeding varieties for drought resistance as well as for sodic soils and tannery effluent affected soils is one of important objective

Breeding methods

1. Introduction
2. Selection
3. Hybridization and selection
4. Heterosis breeding
5. Mutation breeding

List of some varieties/hybrids released in India

Variety	Parentage	Special traits
PPR 2700 (Vakula)	KM 55 × U 22/B	Semi dwarf plants, light purple pigmentation at leaf sheath, compact ears resistant to leaf blast and tolerant to drought in Andhra Pradesh
VR 936 (Hima)	IE 2695 × Godavari (PR 202)	100 – 115 cm tall plants, medium bold, dull white coloured grains purple pigmentation on all parts of the plant
VR 847 (Srichaitanya)	GPU 26 × L 5	Erect plants, medium tall, semi compact ears with tip incurved fingers dark green vegetation, broad fingers
VR 762 (Bharathi)	pure line selection fromVMEC 134	Medium tall plants, incurved fingers, reddish brown colour grains longer intermodal length
OEB 10 (Chilika)	GE 68 × GE 156	85 – 105 cm tall plants, broad and tip incurved (6-8) fingers, light brown colour grains
BM 9-1 (Bhairabi)	Mutant of budha mandia	Semi dwarf plants (85-100 cm), palegreen leaves, tip incurved fingers, light brown colour grains
MR 6 (Divya)	African white × ROH 2	100 – 110 cm tall plants, open ears with tip incurved fingers, average grain yield (q/ha): 30 – 35
PR 230 (Maruthi)	-	Medium tall plants, incurved and medium sized ears, resistant to blast and blight
VR 708 (Champavathi)	Pure line selection from VMEC 36	Purple pigmentation on all parts of the plants, highly susceptible to blast, photo insensitive and early duration
PR 2614 (Saptagiri)	MR 1 × Kalyani	Medium tall plants, semi compact ears with tip incurved fingers, resistant to blast, tolerant to drought
PPR 2350 (Padmavathi)	Pure line selection from Thirupati local	Early duration, suitable for late kharif
PR 1158-9 (Goutami)	Godavari × U 22	Resistant to blast
Maruti (RR 230)	Selection from local collection from Ranaga reddy district	Medium tall with pigmentation at nodes bearing in curved earheads. Non lodging, high yielding variety.

Economic importance: Staple food in parts of East and Central Africa and India. And also it is used for malting and brewing.

□□□

Chapter-7

Breeding in Pulses: Pigeonpea, Chickpea, Black Gram, Green Gram, Cowpea and Soyabean

Pigeonpea (2n = 22)

Pigeonpea or Red gram (*Cajanus cajan*) is one of the major pulse crop of tropics and sub-tropics and cultivated in Asia, Africa, and Latin America. It is consumed on a large scale mainly in south Asia and is a major source of protein for the population of that subcontinent. The genus to which it belongs is *Cajanus* and the tribe *Phaseoleae* of the family *Fabaceae*. Center of origin of pigeonpea is India and Africa. The progenitor of *Cajanus cajan is Cajanus cajanifolius*.

Wild relatives: The genus *Cajanus* is composed of 32-34 taxa divided into three clades: Indian, Australian and Scarabaeoides. No wild conspecific to the cultivated species exists, and thus there are no wild taxa falling within the primary genepool of pigeonpea as defined by Harlan and Dewet (1971). The secondary genepool is comprised of *Cajanus cajanifolius* in the Indian clade, the putative progenitor of the crop, *C. acutifolius*, *C. albicans*, *C. cinereus*, *C. confertiflorus*, *C. lanceolatus*, *C. latisepalus*, *C. lineatus*, *C. reticulatus*, *C. scarabaeoides*, *C. sericeus* and *C. trinervius*. The tertiary gene pool comprises of three additional species namely *C. crassus*, *C. mollis*, and *C. platycarpus*. Majority of these species are endemic and reported in Australia. And therefore, Australia is considered as the centre of diversity for pigeonpea. Wild species *C. lineatus*, *C. sericeus*, and *C. trinervius*, as well as *C. crassus* occurring in areas of notably high annual and/ or seasonal precipitation included the central and southern Indian. Species *C. latisepalus*, *C. cinereus*, *C. acutifolius*, and *C. lanceolatus*, as well as *C. sericeus* and *C. lineatus* encountered in regions of very low annual and/or seasonal precipitation included those from Australian.

Species such as *C. latisepalus*, *C. cinereus*, *C. acutifolius*, *C. lanceolatus* *C. platycarpus* and *C. cajanifolius* are found in high temperature regions of Australian. Populations of northern species included *C. mollis*, *C. trinervius*, *C. confertiflorus*, and again *C. platycarpus* found in low temperature regions. The only one wild relative *C. kerstingii* is reported to occur in West Africa. In addition, *C. scarabaeoides* has also been found in Africa, but spread is restricted to the coastal areas only.

Floral biology: Inflorescence is axillary or terminal racemes on long peduncle. Flowers are papilionaceous, bracteate, ebracteolate; flowers occur in pairs in the axil of each bract, pedicellate, bisexual, complete, zygomorphic, cyclic, hypogynous, pentamerous, dichlamydeous and heterochlamydeous.

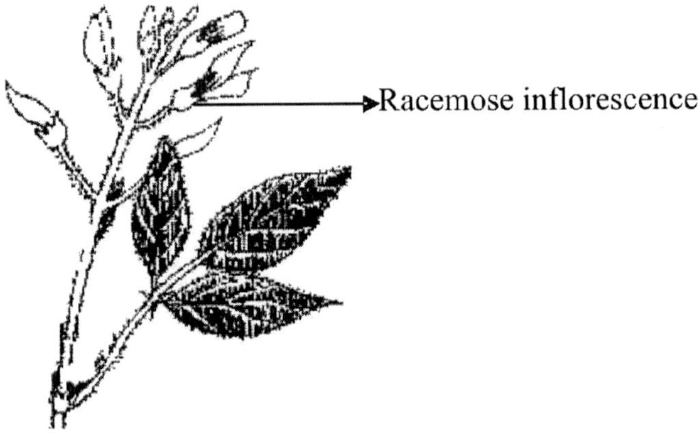

Calyx with five sepals, gamosepalous, green in colour with valvate aestivation. Corolla with five petals, polypetalous, papilionaceous, one vexilium (standard) is quirilled at base, two alae (wing) and two keel petals which are fused at the base, descendingly imbricate aestivation, yellow in colour and blotched with veins or diffuse red.

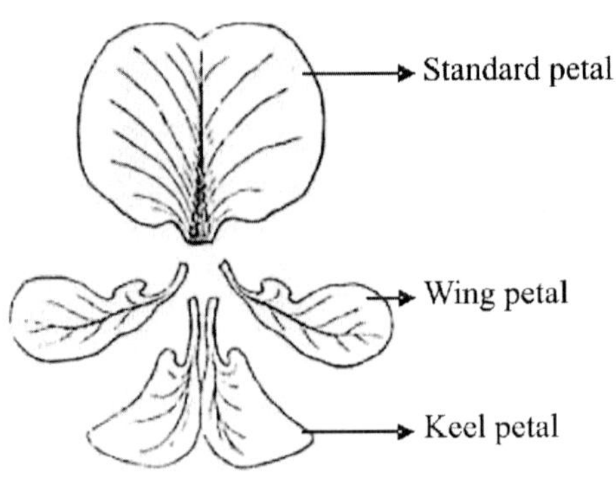

Androeeium consists of 10 stamens which may united in two bundles (9+1), diadelphous nine filaments are fused which are alternately long and short. Anthers are dithecous, introse and dorsifixed. Gynoeeium is monocarpellary, unilocular, superior ovary, with many ovules on marginal placentation. Style is curved (beaked type) stigma capitate and bearded (hairs on one side). A cylindrical membranous disc is present around the ovary. Fruit is a legume, lomentous pod, villous with diagonal furrows. It is variable in size shape, constriction, texture and pubescence. Pods with deep constrictions are known as bearded while others are flattish. Seed colour ranges from cream to black with different shades of yellow, red and brown. A dark chocolate red is the most common colour with a white hilum. Cotyledons are elliptical and deep to light yellow in colour.

Floral formula: Br, eBrl, %, ⚥, $K_{(5)}$, $C_{1+2+(2)}$ A_{9+1}, $\underline{G}_{(1)}$

Floral diagram:

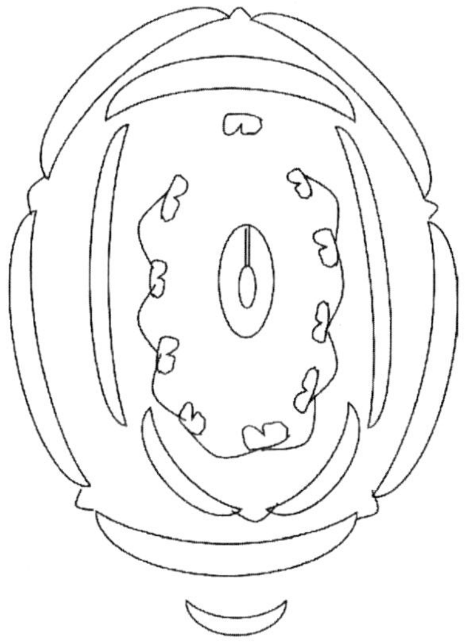

Anthesis and pollination: In pigeonpea, in a fully developed bud, anthers surround the stigma and dehisce a day before the flower opens. Flowers open at about 3-4 P.M and remain open till 2 P.M. The duration of flower opening depends on the climate and environment. Fertilization occurs on the day of pollination. Pigeonpea is an often cross-pollinated crop. The extent of cross-pollination ranges from 3 to 40% with an average of 20%. A plant produces many flowers of which only 10% set pods. For selfing, mature flower buds are to be covered with paper bags for one or two days. Emasculation is by hand emasculation either through ring cut method or keel rupturing method. Young buds which are to open after two days should be preferred for emasculation. Emasculation should be done in the previous day evening and the emasculated buds are protected by covers. Early morning on the next day, pollination is done using pollen collected from the protected flowers of the selected male parents,

Breeding objectives

1. Yield: Breeding for high yielding long duration varieties suitable for rainfed conditions, short duration varieties for irrigated conditions and perennial types suitable for bund cropping needs to be developed.
2. Quality: Average protein content in redgram is 23%. Red grain contains lesser amount of sulphur containing amino acids. So need to increase the protein content along with increasing the amount of amino acids. Red seed coat contains more polyphenol (Tannin) than white seed coat. So preference is towards white seed coat.
3. Resistance to biotic stresses: Among diseases Sterility mosaic, *fusarium* wilt, root rot and stem blight are major diseases. Among pests, *Helicoverpa* borer and pod fly are the major pests; terminal cluster types are highly susceptible to this pest.
4. Resistance to abiotic stresses: In this crop, adverse conditions like drought and water logging causes maximum yield losses throughout semi arid tropics.
5. Breeding for vegetable type: Pigeon pea seeds when used as green as a vegetable are nutritionally better than mature seeds as they have protein, sugar and fat content.
6. Breeding for early maturity: Needs to develop short duration varieties suitable for multiple cropping/intensive cropping and also for escape from abiotic stresses like terminal drought etc.

Breeding methods

1. Introduction: The varieties developed and released in other states or counties imported and after evaluation may be released for commercial cultivation.
2. Selection: Pure line selection was used in the development of varieties N-148, BDN-1, BDN-2 etc.

3. Hybridization and selection: Pedigree method was used in the development of varieties like T-21, ICPL87 (pragati) and ICPL 8863.
4. Mutation breeding: ICPL-270, Vishal 1 and TA10 are the varieties resulting from mutations only.
5. Population improvement using recurrent selection methods is also possible in redgram.
6. Transgenic breeding: Development of transgenics for biotic stresses, insect pests and abiotic stresses like drought, water logging, salinity and thermo-insensitivity and also herbicide resistance are important targets.

Hybrid pigeonpea: The hybrid technology is based on three major components namely, male sterility and its genetic maintenance system, stable fertility restoration, and cost effective mass pollination mechanism. In pigeonpea, so far three forms of male sterility systems differing in their genetic control have been reported. These include genetic male sterility (GMS), temperature - sensitive male sterility (TGMS), and cytoplasmic nuclear male sterility (CMS). The GMS was the first to be discovered in 1978 but it could not be used for large scale hybrid seed production due to the need of maintaining male sterility through heterozygotes. The TGMS is of recent origin and needs a lot of research before it could be used in commercial hybrid breeding. The third form of male sterility, CMS, was developed through a cross involving a wild relative of pigeonpea (*C. cajanifolius*) as female parent with a cultivated line as male parent and it is the most valuable product as far as commercial hybrid breeding is concern. Hybrid ICPH 8 was bred by hybridizing a genetic male-sterile line ms Prabhat (DT) and a male-fertile inbred line ICPL 161, both developed at ICRISAT Center and it is the irst hybrid of a pulse crop released for commercial cultivation. Recently three more hybrids released in India are ICPH 2671 in 2010, ICPH 3762 in Odisha in 2014 and ICPH 2740 in Telangana. These three are based on cytoplasmic male sterility only.

List of some varieties/hybrids released in India

Variety	Parentage	Special traits
ICP 8863	Selection from ICP 7626	*Fusarium* wilt resistant high yielding, medium duration, suitable for sole and intercropping
UPAS -120	Selection from P 4758	Extra short duration variety and yield 1.5-18 t/ha
PDM 1	Selection from peddapuram local	Good in cooking quality. Suitable for high rainfall areas.
Palnadu (LRG 30)	Pure line selection from local 'Jada kandi'	High yielding variety with wide adaptability
CO 1	Local (Thamilanadu)	Photosensitive brown seeded, average yield – 1.5 t/ha
AS71-37	Local selection	Medium tall potential yields 3t/ ha,
BDN 1	Local selection	Semi spreading, wilt resistant
HY 5	PL 3701	Semi erect , purple stem medium height (170 cm)
AS 71-37	Local selection	Medium tall (172-194 cm), Potential yield (3t/ha)
BDN -7	Local selection	Semi- spreading, wilt resistant.
C11	Local selection	Profuse branching, browning.
Gwalior 3	Local selection	Tall (250-300 cm), light brown seed, 7-8gm (100 seed)
Prabhath	T1 × T 190	Determinate, dwarf clustered fruiting, yield (1.2-1.5 t/ha)
ICPL	T 27 × JA 277	Determined, extra short duration vatiety brown seeded, yield (2.5-3t/ha)
Pragathi	T 27 × JA 277	Determined, extra short duration vatiety brown seeded, yield (2.5-3t/ha)
ICPL 157 (Jagriti)	ICP 6997 × Prabhath	Determinate, yield (2.5-3t/ha), cream colour.
Pant A3	-	Determinate, semi erect
Pusa Ageti	BAJIL 1-7 ×TI 90	Dwarf compact plant, bold seeded 9gm (100 seed)
T 21	TI × TI90	Semi spreading 7gm (100 seed), short duration variety.
Pusa 84	Pusa Ageti × T 27	Medium tall, 7.5gm (100 seed)
S20	Kankel × B7	Semi compact, medium duration vatiety.

Bahar	D1258 × Local	Compact, 4-5 seeds/ pod, Semi erect plant.
Lakshmi (ICPL 85063)	BDN 1 × (T 21 × JA 275)	Bold red seeded with resistant to sterility mosaic virus.
ICPH 2671	ICPA 2043 × ICPR 2671	It is stable for its fertility restoration
ICPH 2740	-	Resistant to wilt and sterility mosaic.
ICPL 14003	-	Resistant to terminal droughty, suitable for light chalka soils.
ICPL 3320WR	-	Tolerant to helicoverpa
Asha (ICPL-87119)	-	Kharif/ semi rabi cultivation in vidarbha under contingent situation
ICPH 3762	-	High yield
ICPH 2740	-	Drought resistant
Ranga bold (LRG 38)	C 11 × ICP 7035-1	Bold seeded variety suitable for sole cropping as well as intercropping
LRG 41	-	Resistant to drought, pod borer, *heliothis*.
LRG 52	-	Moderately tolerant to helicoverpa, pod fly, *fusarium* wilt.
ICPL 332	-	Moderately resistant to pod borer.
MRG 66 (Madhira 66)	-	Resistant to *macrophomina*
PRG 100	RRG 65 × Pusa Ageethi	Early maturity and resistant to wilt
ICPL 84031	-	Extra early maturity suitable for multi-cropping system.
AKPH -4101	GMS based hybrids	Early maturing variety.
Santura -1	-	Mild late maturing variety.
Vamban-1	-	Under rainfed condition yield (840 kg/ha)
Chamatkar (Malviyal)	-	400-600 pods/plant
Erramacha kandi	-	
KPP008	-	Early maturity.

Economic importance: The green pods and green seeds are eated as vegetable. The ripe dry seeds are boiled and eaten as pulse split, pulse, is used as dal. Dried husk and broken pieces of seed are used as cattle feed. Green leaf and tops of the plant is used as fodder and for making silage. Dried stake is used as fuel and as thatching material.

Chickpea (2n = 16)

Chickpea (*Cicer arietinum*) is an important grain legume of the semi arid tropics and warm temperate zones, and forms one of the major components of human diet. Chickpea is cultivated from the Mediterranean basin to the Indian sub-continent and southward to Ethiopia and the East African Highlands. It has been introduced to the Americas and gained popularity especially in Mexico. However, the crop assumed greatest significance in Indian subcontinent. The genus to which it belongs is *Cicer* and the tribe Cicereae of the family *Fabaceae*. Center of origin of chickpea is South East Turkey and Syria. The progenitor of *Cicer arietinum is Cicer reticulatum.*

Wild relatives: The genus *Cicer* contains 43 species and is devided into four sections *Monocicer, Chamaecicer, Polycicer, Acanthacicer.* Together with eight other annual wild relatives, cultivated chickpeas are placed in subgroup *Pseudononis*. The remaining 34 wild *Cicer* species are perennial shrubby plants and comprise subgroup *Viciastrum*. The wild relatives of chickpea can be grouped into primary, secondary and tertiary gene pools, according to their cross-compatability and fertility of interspecific derived hybrids. Primary gene pool consists of species *C. arietinum*, known only for cultivation and *C. reticulatum*, endemic to SE Turkey and weedy wild progenitor. Secondary gene pool consists of species *C. echinospermum*, distributed in Turkey and Iraq. Tertiary gene pool consists of *C. bijugum* distributed in Iraq, Syria and Turkey, *C. judaicum* distributed in Israel, Lebanon, Palestine and Turkey, *C. pinnatifidum* distributed in Cyprus, Iraq, Turkey, Syria, Lebanon and Armenia, *C. cuneatum* distributed in North and North east Africa, *C. yamashitae* endemic to Afghanistan, *C. chorassanicum* distributed in Afghanistan and Iran.

Cicer cultivated grains are classified into two groups:

Desi or brown type: Most widely grown, seed colour ranges from yellow to dark reddish brown.

Kabuli or white type: Distinct species, seeds bold, yield comparatively poor.

Floral biology, floral formula and floral diagram is same as like of pigeonpea.

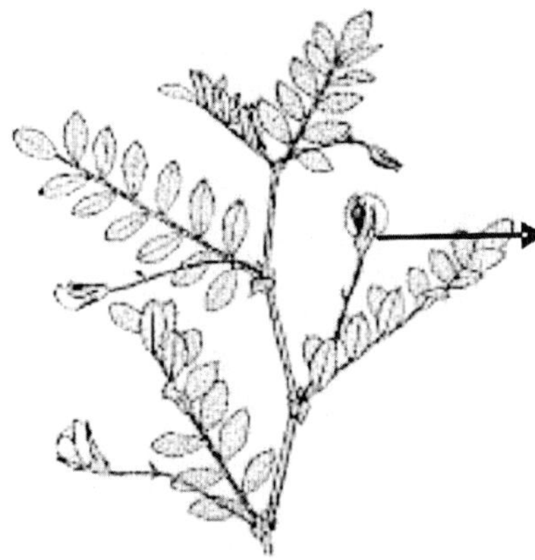

Anthesis and Pollination: Anther dehiscence takes place inside the bud one day before the opening of the flower. When pollen are first liberated, the stigma is still above and quite free from the base of the anthers. The filament gradually elongates to carry the anthers above the stigma. This process is completed before the flower opens, thus facilitating self-pollination. The flower open in the morning and close in the afternoon and each flower opens on two or three successive days. Time of anthesis is 3 AM to 9 AM. For selfing, mature flower buds are to be covered with paper bags for one or two days. For hybridization crossing work should be started when the first pod on the selected plant is already formed. Emasculation is by hand emasculation either through ring cut method or keel rupturing method. In Northern India, emasculation is done a day prior to pollination. The pollination is done in the morning hours give better setting. In south India, pollination immediately after emasculation gives higher seed setting.

Breeding objectives

1. Yield: Along with high yield improvement, stabilization breeding was considered to be important significance in this crop.
2. Quality: Quality in chickpea is mainly determined by grain size colour, cooking quality and nutritive value.
3. Resistance to biotic stresses: In chickpea, blight and wilt are major diseases. *Helicoverpa* pest is the major problem and also herbicide resistance was poor in this crop.
4. Resistance to abiotic stresses: Drought, heat, frost, saline and alkaline are the major problems in this crop.

5. Resistance to shattering: Ripening of pods in some vareties tends to dehisce and release the seeds. In order to achieve high yield, the varieties will not shatter even after maturity.

Breeding methods

1. Selection: Wilt resistant variety JG315 was developed through selection from local variety.
2. Hybridization and selection: Bulk, pedigree and mass pedigree methods have been employed in the segregating generations for the development of high yielding varieties. N-31, N-59, Vikas, Vishal, Vishwas and G-12 are the varieties released based on pedigree method.
3. Transgenic breeding: Development of transgenics for resistance to drought, frost as well as for *Helicoverpa* pest is the major recommendation.
4. Mutation breeding, use of somaclonal variations and heterosis breeding are some of the other breeding methods commonly used in chickpea breeding.

List of some varieties/hybrids released in India

Variety	Pedigree	Salient character
ICCC 32	-	Tolerant to root and pod bores
Kranti (ICCC 37)	P 481 × (JG 62 × P1630)	Semierect with medium, bold, round attractive brown seeds. Tolerant to fusarium wilt and dry root rot diseases.
Sweta (ICCV 2)	(LK 850 × GW 7) × (L550 × Guamuchli 2)	A kabuli, semi spreading and early maturing type resistant to fusarium wilt.
Pusa 256(BG256)	(JG62 × 850- 3/27) × (1550) × H208	Medium tall, leaf broad light brown seeds, tolerant to *Ascochyta* blight and wilt
Pusa417(Grinas/ BGM417)	Induced mutant of BG203	Suitable to normal & late sown condition
ICCC 37 kranti	(P 481 × JG 62) × P 1630	Dwarf leaves, light green narrow semi erect seeds
Pusa 372 (BG 329)	P 1231 × P1265	Semi spreading, resistant to wilt, blight and root rot, suitable for late sown conditions

JG 218	ICCC4 × P1353	Early maturity tolerant to wilt
DCP 92-3	Selection from GP	-
JGG1	Selection from land races of gulabi gram	Suitable for table purpose and good punching quality
Gujarat Junagadh gram 1 (GCP101)	JG1 × ICCCV 2	-
Gujarat Junagadh gram 2 (GCP107)	JG1258 × BDN 9-3	-
RSG 895 (Arpita)	RSG 44 × RSG 256	Resistant to dry root rot and wilt
RSG 945 (Asha)	RSG 668 × RSG 617	Resistant to wilt and nematode.
Pusa shubra (BGD 128)	ICCV2 × ICCV5	Semi erect, light green foliage low seed Weight 27.7 grams

Economic importance: It is widely cultivated for its nutritious seeds, which are harvested when immature and eaten raw, roasted, or boiled or when mature and dry processed into flour.

Black gram/urd bean (*Vigna mungo*) (2n = 22)

Green gram/ mung bean (*Vigna radiata*) (2n = 22)

Cow pea (*Vigna unguiculata*) (2n = 22)

The genus *Vigna* is a large and immensely variable genus comprising more than 200 species. This genus contains several agronomically important species that have considerable economic and environmental importance. These include the mung bean (*V. radiata*), urd bean (*V. mungo*), cowpea (*V. unguiculata)*, azuki bean (*V. angularis*), bambara groundnut (*V. anpurean*), moth bean (*V. aconitifolia*) and rice bean (*V. anpure*). Most of these vignas are grown in the warm temperate and tropical regions of the world. The major areas producing these crops include countries from Asia, Australia, West Indies, South and North America, and tropical and subtropical Africa. The genus *Vigna is* belongs to the family *Fabaceae*.

Mung bean (*Vigna radiata*) and urd bean (*Vigna mungo*) are most important *vigna* crops and are believed to have originated in Indian subcontinent. The progenitors of mung bean *Vigna radiate* var. *sublobata* and urd bean *Vigna munga* var. *sylvestris* are seen in abundance as weeds in India as well as in wetlands of subtropical; regions of northern and eastern Australia. The origin and domestication of cowpea (*V. unguiculata*) is believed to have occurred in West or Central Africa. For rice bean (*V. anpure*), India is the main center of origin and var. *gracilis* is likely to be an ancestor of rice bean. Another wild species *Vigna minima* was considered as a wild relative of rice bean which has been found in India. Moth bean (*V. aconitifolia*) was considered as one of the primitive species of *vigna* originated in India center of origin. Similarly azuki bean (*V. angularis*) originated in Chinese center of origin and wild ancestor of cultivated azuki bean is *V. angularis* var. *nipponensis*. In below table shows various vigna species are grouped in to primary, secondary and tertiary gene pools on the basis of *V. anpur* and *V. mungo* put in primary gene pool.

Crop	Primary gene pool	Secondary gene pool	Tertiary gene pool
Mung bean	*Vigna anpur* *V. anpur* var. *anpur* *V. anpur* var. *sublobata* *V. anpur* var. *setulosa*	*V. mungo* var. *mungo* *V. mungo* var. *sylvestris* *V. aconitifolia* *V. trilobata*	*V. angularis* *V. dalzelliana* *V. glabrescens* *V. grandis* *V. umbellate* *V. vexillata*
Urd bean	*V. mungo* var. *mungo* *V. mungo* var. *sylvestris*	*V. anpur* var. *anpur* *V. anpur* var. *sublobata* *V. anpur* var. *setulosa* *V. aconitifolia* *V. trilobata*	*V. angularis* *V. dalzelliana* *V. glabrescens* *V. grandis* *V. umbellate* *V. vexillata*

Flower is axillary or terminal racemose clusters, 5-6 flowered on swollen nodose peduncle. Floral biology, floral formula and floral diagram is same as like of pigeonpea.

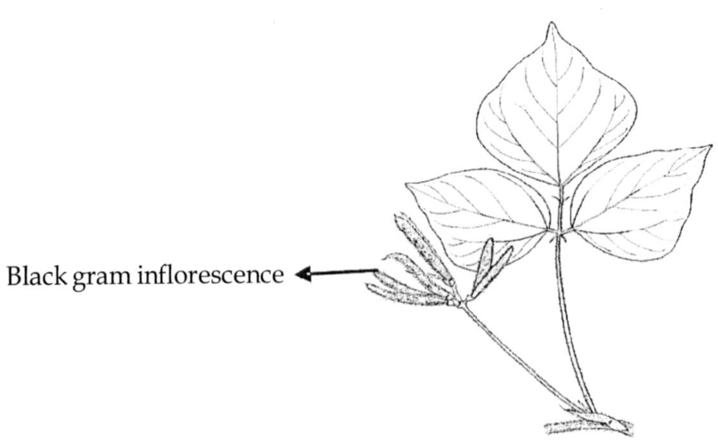

Anthesis and pollination: In these pulse crops, self pollination is the rule because pollination occurs before flower opening (cleistogamous) in night. The flower opens in the morning between 6AM to 7 AM. Cleistogamy is recorded to an extent of 42%. The interval between pollination and opening of flower is 4 hours. This ensures self fertilization. For selfing, mature flower buds are to be covered with paper bags for one or two days. Emasculation is by hand emasculation either through ring cut method or keel rupturing method. Young unopened bud is kept between thumb and fore fingers of the left hand. The point of dissecting needle is inserted just under the standard petal in an oblique position along the top of the bud. The left side of the standard and wing petal are pushed outward and held with thumb and left hand. The left side of the keel petal is removed with the forceps. The pistil and stigma are then exposed and the anthers are removed with the forceps. Evening emasculation followed by morning pollination gives best results. Pollination is done by gently rubbing anther of male, inserting the staminal column and closing it with standard and wing petal. Since flower shedding is common, putting better paper bag is avoided. The emasculated flowers are identified with thread wound round. The crossed pod will be smaller in size with two or three seeds only.

Breeding objectives

1. Breeding for high yielding medium duration varieties for dry land cultivation, high yielding varieties short duration suitable for irrigated conditions or rice fallow situations are main targets.
2. Quality: It is determined by protein content, methionine content, cooking quality, percentage of hard seeds and dal recovery percentage etc.
3. Resistance to biotic stress: Yellow vein mosaic virus (YMV) was identified as the most critical yield limiting problem followed by powdery mildew and leaf spot are the major problems. Among the pests, White fly vector for YMV and leaf crinkle, leaf eating caterpillar are major problems.
4. Other includes problem related to sprouting of seeds *in situ* under rains, photo-insensitiveness, resistance to abiotic stresses like drought, early maturity are important breeding objectives.

Breeding methods

1. Introduction: The varieties developed and released in other states or counties imported and after evaluation may be released for commercial cultivation.
2. Selection: Pure line selection is commonly practiced
3. Hybridization and selection: Inter varietal and inter specific hybridization methods used for transfer of disease resistance as well as for transfer of methionine content from black gram to green gram
4. Mutation breeding: Natural mutation or induced mutation can also be used in development of improved varieties.
5. Transgenic breeding: Development of transgenics for YMV resistance through pathogen derived resistance is the important goal. And also genetic engineering for resistance to pre-harvest sprouting was another important goal identified in this crop.

List of some varieties/hybrids of black gram released in India

Variety	Parentage	Special traits
ADT-1	Selection from aduthurial local	High protein content (19.9%)
Krishnayya (LBG 17)	Netiminumu × Chikkudu minumu	For rice fallows in rabi. Resistant to powdery mildew.
Kottabutta minumu (Lam611)	Pure line selection from local butta minumu	Erect habit with vigorous initial growth with profuse branching.
LBG 648	Krishnayya × Butta minumu	Photo insensitive and suitable for rice fallow situations
Warangal 26	T9 × Line 400	High yielding variety with new plant type-sympodial bearing habit.
Podalakur minumu (PBG 1)	Pure line selection	Photo insensitive bearing pods in clusters suitable for rice fallows.
CO-1	Pureline selection from srivaikundam local	Released for rainfed conditions
CO-2	Pureline selection from PLS 150	Photo insensitive
CO-3	Pureline selection from PLS-364	Less susceptible to Root rot & pod borer

KM1	G31×kargon 3	Dwarf type & drought tolerant
CO-4	Induced mutant from CO-1	Tolerant to leaf crinkle ,virus,tip blight, powdery mildew& less susceptible to stemfly & pod borer
KM2	T9×L64	Dwarf type & tolerant to viral diaeaaes
ADT-2	Derivative of tirunelvili× ADT-1	High protein content (21.5%)
TMV-1	Midhi ulundu×KM-1	Resistent to YMV & tolearant to root rot
CO-5	Pureline selwction from musiri type	Moderately resistant to powdery mildew,leaf crinckle,pod borer & tip blight
ADT-3	Pureline selection from tirunelveli	Suitable for rice follows
UG-218	T9×LU 220	Resistant to YMV & dwarf plant
LBG-17	-	Rice follows and resistant to YMV
ADT-4	T9/ADT2/PANT U-14 cross derivatives	Suitable for raising in field bunds, higher grain weight,resistant to Ymv & stem fly
Vamban-1	KM-1×H 76-1	HYV & suitable for entire state & tolerant to YMV
ADT-5	Pureline selection from anpur variety	Suitable for rice follows,resistant major diseases like YMV,root rot & leaf crinckle & stem fly
LBG 402	-	Resistant to wilt
PDU-1	Selection from IC 8219	Tolerant to YMV & dwarf plant and matures in 70-80 days
TMU	AB 133× ADT 1	Suitable for tamilnadu region
APK-1	ADT 2 × RU 1	Suitable for rainfed condirions for intercropping with cotton
K-1	CO-3 × VS 131	Suitable for southern districts ,YMV resist
Vamban -2	Spontaneous mutant selection from typer	Resiatant to YMV
Vamban -3	LBG 402 × LBG 17	Resiatant to YMV
LBG – 645	-	HYV , wilt resistant & bold shining seed
Vamban-4	CO4×PDU102	resistant to yellow mosaic virus
Vamban-5	Vamban1×UK17	HYV & resistant to YMV
IPU02-43	DPU88-31×DUR1	Resistant to YMV & powdery mildew & short duration variety

TNAU-8	VBN3×VBN4	Non- shattering & synchronous maturity & resistant to leaf crinkle
VBN-6	Vamban1×vigna mungosilvestris	Resistant to YMV & early harvest
TNAU-7	Vamban3×vignasivestris	Resistant to YMV& synchronous maturity

Black gram economic importance: Seed is used as a source of protein (23.4%). The cotyledon contains batter (aminoacid related substance). This makes the wet flour to swell up i.e. doughness. Because of this batter, it is suitable for making idlis, dosa etc.

List of some varieties/hybrids of green gram released in India

Variety	Parentage	Specific traits
Type-44	Type 1 × Type49	Resistant to YMV
Pushkara (LGG 450)	Gamma radiation mutant from ML 26-10-3	Suitable for all seasons, recommended for rice fallows.
LGG 407	Gamma radiation mutant from 26-10-3	Resistant to yellow mosaic virus, foot rot and angular leaf spot disease.
LGG 460	Lam M2 × ML 267	Resistant to yellow mosaic virus, photo insensitive and high yielding variety.
Bhadradri (MGG- 348)	MGG 329 × NARP 1	Tolerant to drought multiple resistant culture to YMV, leaf crinkle virus, leaf spot. Dwarf suitable for intercropping
KM2	Hybrid derivative of 127 × S9	-
Jyothi (HUB 4-3)	L-24-2 × Pusa bhqisaki	Resistant to YMV
Pusa-101	(SB × EG-MG-16) × MG5010A(Y)ML5	Resistant to LS and YMV
Pant mung 4	TYPE 44×UPU2	-
ML-337	ML1 × AM987	Resistant to LS, YMV and PM
Pant mung3	LM294-1×L-80	Resistant to PM ,YMV
K1	CO4 × ML65	-
ADT-3	H7016 × Rajendra	Suitable for only Rice fallow
VBN-1	Hybrid derivative of S8 × PIM S3	-
CO6	WGG37 × CO5	–

HUM-1	PHUM1 × Pant U30	-
VBN(Gg)2	VGG4 × MH309	Trifoliate with lobes sparsly hairy pods shiny green
GM-4	GM3× Pusa 9333	-
CO(Gg)7	MGG336 × COGG902	High protein content synchronous maturity
VBN(Gg)3	CO1 × Vellore local	Moderately Resistant to PM
TM-2000-2	TARM-1 × J781	Resistant to PM
IPM99-125	Pantmung2 × AMP36	-

YMV- Yellow mosaic virus, LS – Leaf spot, PM – Powdery mildew

Green gram economic importance: Mungbean has highest digestability among the grain legumes. It is free from causing flatulence, mostly used as sprouts, also as split dal. The rich source of protein (23.5%).

Soya bean (2n = 40)

Soya bean (*Glycine max*) is the world's most important seed legume, which contributes to 25 % of the global edible oil, about two-thirds of the world's protein concentrate for livestock feeding. The cultivation and use of soybean could be traced back to the beginning of China's agricultural age. Today, USA, Brazil, and Argentina are the "Big-3" producers of the world. In India, Soybean was introduced from China in tenth century AD through the Himalayan routes, and also brought in via Burma (now Myanmar) by traders from Indonesia. The genus to which soya bean belongs is *Glycine* and the tribe *Phaseoleae* of the family *Fabaceae*. Center of origin of soya bean is China. The progenitor of *Glycine max is Glysine usuriensis.*

Wild relatives: The genus *Glycine* includes about 20 annual and perennial species, distributed for the most part in Australia and Asia. It is divided into two subgenera, *Soja* (annual) and *Glycine* (perennial). Domesticated soybeans (*Glycine max*) are included in subgenera *Soja,* together with their closest wild relatives, which are found in East Asia (China, Korea, Japan and Far East region of Russian federation). Subgenera *Glycine* is comprised of the roughly 15 perennial wild relatives of soybeans. The majority is both native to and restricted to Australia, but the geographic range of some species extends to SE Asia, and even as far north as Japan. These perennial wild soybean relatives mainly come under tertiary gene pool system. The taxonomic boundaries of the wild soybean (*Glycine soja*) complex have not been conclusively resolved, and one to three closely related, interfertile wild relatives (*G. soja, G. gracilic* and *G. formosana*) have been delineated.

Floral biology: Inflorescence is Axillary raceme bearing 3-15 flowers. Flowers are white, small, bracteate, ebracteolate, pedicellate, complete, bisexual, regularly zygomorphic, pentamerous cyclic, dichlamydeous and heterochlamydeous.

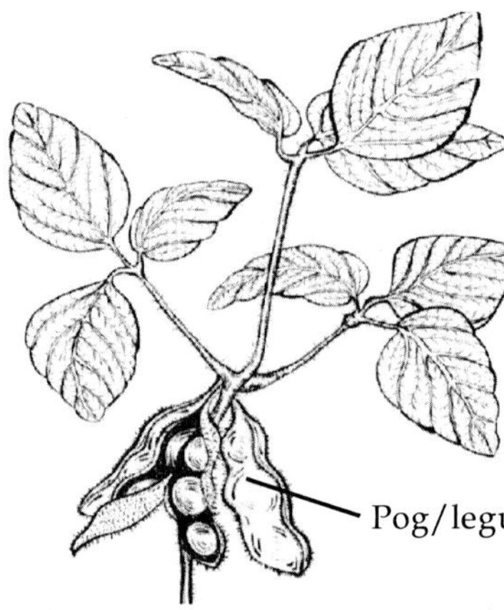

Calyx consists of five sepals, gamosepalous, persistent, anterior tooth longest, united at base and free above with valvate aestivation. Corolla consists of five petals, papilionaceous, polypetalous, descendingly imbricate aestivation. Androecium consists of ten stamens with monodelphous, globbose, dithecous, introrse and dorsifixed anthers. Gynoecium with monocarpellary, unilocular, superior ovary with 1-5 ovules on marginal placentation, style is curved, gabrous with capitates stigma. Fruit is a legume or pod, borne in clusters, profusely hairy, dehiscent, linear, oblong recurved. Seed is globular, dull white or cream or black dicotyledonous non-endospermic.

Floral formula and floral diagram are similar as like Groundnut.

Anthesis and pollination: Flower opens early in the morning. The pollen is shed normally shortly before or after the flower opens. But pollen shedding may occur sometimes with in the bud itself. Normally cross pollination does not exceed 1 percent. Hand emasculation is the method followed for crop breeding which is tedious since the floral parts are so small and seed set is also less. Emasculation is done in the evening and pollination is done in the morning hours.

Breeding objectives

1. Yield: Breeding high yielding short duration varieties is the important target. Yield in soy bean mainly depends on size, number of seeds per pod and number or pods/plant.
2. Quality: It mainly depends on Seed color and quality, oil content and quality, protein content and seed coat color. Protein content is negatively correlated with oil content so white breeding for high protein content a compromise is to be made.
3. Breeding for biotic stress resistance: Among the diseases, yellow mosaic virus is considered to be important problem, besides this, soy bean rust, root rot, pod blight and bacterial blight were considered noteworthy. Among the pests, stem fly, stem borer and green semi looper were identified as major problems.
4. Breeding for non shattering, lodging resistant, breeding for vegetable type, forage type and varieties for rice follow situation are the other important objectives.

Breeding methods

1. Introduction
2. Selection: Pure line selection is commonly practiced in soy bean.
3. Hybridization and selection: Inter varietal and inter specific breeding methods are commonly practiced.
4. Mutation breeding
5. Transgenic breeding: Development of transgenics for resistance to YMV and fungal diseases is important criteria in soy bean.

List of some varieties/hybrids released in India

Variety	Parentage	Specific traits
Tams 38(ICAR) (IC295927)	Monette × PK 472	Early maturity determinate growth,15-20%
LSB-1	Selection from MACS 330	Extra early suitable for intercropping in cotton and red gram.
TAMS-98-21 (IC -415096)	Mutant of JS-8D-21	21% more yield High PUFA(59%)
Kalitur	Indigenous variety	Semi determinate
Bragg	Introduction from USA	Determinate growth
Improved pelican	Introduction from USA	Semi determinate growth

Hardee	Introduction from USA	Determinate growth
Punjab 1	Selection from Nanking	Semi determination growth
Durga	EC 14437 × Bragg	Semi determinate growth
PUSA 40	8-3 × Lee	Determinate
PK 308	T 31 × Hardee	Determinate growth
PK 471	Hardee × Punjab	Determinate growth
JS 71-05	Selection from exotic material	Determinate growth
JS 335	JS 78 – 77 × JS 71-5	Determinate growth
NRC 2 (Ahilya 1)	Mutant of Bragg	Determinate growth
PS 1042	Bragg × PK 416	Determinate growth
NRC 37 (Ahilya 4)	Gaurav × Punjab 1	Semi determinate growth
JS 93 - 05	Selection from PS-73-22	Determinate growth
JS 95-60	Selection from PS 7322	Determinate growth
PS 1347	PS 1024 × PK 472	Semi determination growth
SO -1	PHAS loci mi R4 392 gene	High protein content
SO -3	ArrpphB gene	Resistant to Soyabean Mosaic Virus (SMV)
SO -4	Rps gene at QTL loci	*Phytopthora sojea* resistant variety
SO -6	Rpp 1 gene producing Myb 84 transcription factor	Soya bean rust resistant
SO -7	IDA genes	Resistant against root knot nematodes
S - 39	Mutation of CHS7 & CHS8 genes	Yellow seed coat colour with high oil content
SO - 41	Two way cross (RIL × Williams -82) x *G soja*	Hard seededness
S -35	Rhg1 loci of soya bean germplasm	Soya bean cyst nematode resistant
S - 36	Mutation of S– 35 using N-nitrose N–methylurea (NMU)	Increases metabolisable energy (ME) during processing of soya bean meal
S -22	RNA I technology of traditional varieties	Reduce in B–conglycinin content in oil cake to make it suitable for cattle feed.

Economic importance: Seeds have high protein (40%) and oil content (20%). Black seeds are rich in protein and low in oil, yellow seeds are low in protein and rich in oil.

❑❑❑

Chapter-8

Breeding in Oilseeds: Groundnut, Castor, Sesame, Sunflower, Safflower, Niger and Mustard

Groundnut (2n = 40)

Groundnut (*Arachis hypogaea*) also commonly known as peanut, is important oilseed crop, grown throughout the tropics and subtropics. Groundnut predominantly grown in Asia, which accounts for 64% of the world production and in Africa, which accounts for 26% of the world production. More than 25% of the groundnut area harvested in the world is in India followed by in China. Its seeds are a rich source of edible oil (43–55%) and protein (25–28%). The genus to which groundnut belongs is *Arachis* and the tribe *Aeschynomeneae* of the family *Fabaceae*. Center of origin of groundnut is Brazil. The progenitor of Arachis *hypogaea is Arachis monticola.*

Wild relatives: The genus *Arachis* has been divided into nine sections i.e. *Arachis, Caulorrhizae, Erectoides, Extranervosa, Heteranthae, Procumbentes, Rhizomatosae, Trierectoides* and *Triseminatae*. Among these, the section *Trisectoides* is considered to have the most ancestral characters, such as tuberous hypocotyls or roots, trifoliated leaves and vaginated stipules. And sections *Extranervosa, Heteranthae* and *Triseminatae* are most primitive and that sections *Caulorrhizae, Erectoides, Procumbentes, Rhizomatosae* and *Trierectoides* are in intermediate position. The section *Arachis* is considered to be the most diverse and derived harbouring both annual and perennial diploid (2n = 20) and tetraploid (2n = 4x = 40) species. The cultivated groundnut, *Arachis hypogaea* comprises two subspecies *hypogaea* and *fastigata*. These in turn, have two (*hypogaea* and *hirsuta*) and four (*fastigiata, vulgaris, aequatoriana* and *peruviana*) botanical varieties respectively.

Species	*Arachis hypogaea*					
Subspecies	*hypogaea*		*fastgiata*			
Botanical varieties	*hypogaea*	*hirsuta*	*fastigiata*	*vulgaris*	*aequatoriana*	*peruviana*
Agronomic types	Virginia/ Runner	Peruvian Runner	Valencia	Spanish		

The type variety *Arachis hypogaea* subsp. *hypogaea* var. *hypogaea* has a long life cycle, no flowers on the central stem and regularly alternating vegetative and reproductive side stems. It is widely present as landraces along the tributes to the south of the Amazon River in Brazil and Bolivia. The modern agricultural types Virginia or runner exemplify this type. Another variety hirsute (Peruvian runner) with more hirsute leaflets and even longer life cycle and is concentrated in the coastal regions of Peru, from where it extends to Central America and Mexico, Asia and Madagascar. The subspecies *fastiagata* has a shorter cycle, flowers on the central stem and reproductive and vegetative stems distributed in disorganized way. The variety *vulgaris* has its distribution centred on the basin of the river Uruguay and it corresponds to the agricultural type known as Spanish. The variety *fastiagata* has its center of diversity in Paraguay and central and north eastern Brazil extending to Peru and it corresponding to the agronomic type known as Valencia. The other two varieties *aequatoriana* in Ecuador and North of Peru and *peruviana* in Peru, North east of Bolivia and the Brazilian state of Acre with restricted distribution.

Floral biology: Inflorescence is Axillary, solitary condensed cymes. Sometimes 3-5 flowers arise in each axil. Flower is Bracteate, bracterolate, bracteoles are two in number and narrow membranous fuse to form a single structure, the tip is bilobed, complete perigynous, flowers sessile or pedicel is very small, regularly zygomorphic, bisexual, cyclic pentamerous, dichlamydeous, heterochlamydeous.

Flower

Calyx with five sepals, gamosepalous, all the sepals are united to form a long tubular structure called hypanthium (Calyx tube which gives the false appearance of a pedicel). A single bract and two bracteoles are present for each flower, the calyx is tubular, slender ending in five lobes of which 3 are united into one big strucuture and the two linear (Lanceolate). Corolla with five petals, polypetalous, papilionaceous, vexillary descendingly imbricate aestivation, yellow in colour, corolla arising at the rim of the calyx tube. Keels are united. Androecium consists of 10 stamens, out of which 8 fertile and 2 staminodes (sterile) monodelphous, the anthers are dimorphic 4 anthers are linear with short filaments and 4 anthers are round with long filaments, sepals, petals and anthers appear to arise from the rim of the calyx tube. Gynoecium with superior ovary, monocarpellary, unilocular with 1-3 ovules on marginal placentation. Ovary situated at the bottom of the calyx tube, the style very slender passes through the length of the calyx tube on the monodelphous staminal bundle, me style is filiform; stigma is button shaped, terminal and hairy below. Fruit is indehiscent pod and inflated legume with 1-3 seeds. Pericarp constricted between seed and reticulates. Seeds are non endospermic, rich in oil and protein. Seed coat is thin papary, rose in colour but varying from dark purple to dark red and white.

Floral formula: Br, eBrl, %, ⚥, $K_{(5)}$, $C_{1+2+(2)}$ A_{10}, $\underline{G}_{(1)}$

Floral diagram:

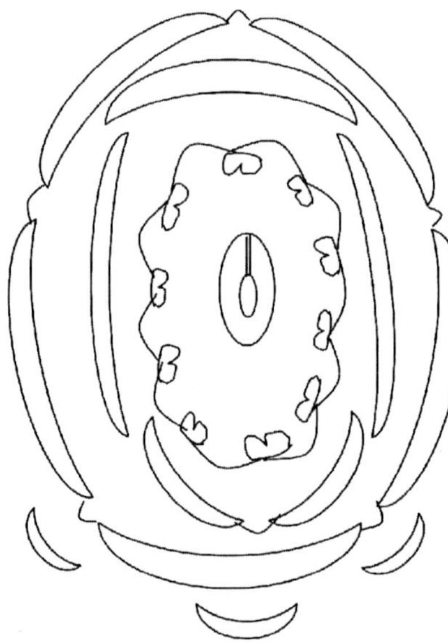

Anthesis and pollination: Self pollination is the rule in groundnut. Anthesis commences at 6 AM and continues upto 8 AM. Anther dehisces two hours prior to opening of the flower. Twenty four hours before anthesis, the buds are very small. During the day, elongation of calyx, proceeds slowly but process gets accelerated during night. Since cleistogamous condition prevails in groundnut, selfing is most easy in this crop. Usually covering is unnecessary and difficult. Keeping the plants in insect proof cages will ensure self pollination. Mature flower buds which are ready to open in the next day are selected and emasculated in the evening. For emasculation, hold the bud between the thumb and the index finger of the left hand and with the help of a razor blade in the right hand; make a cut on the depressed side at two-thirds the length below the tip so as to cut the standard and a portion of the wing petals. Then gently pull the calyx and corolla by holding at the tip of the flower bud. By doing this, the sepals and the petals except the keel would be removed, with the help of the fine forceps gently liberate the bundle of stamens and the pistil from the keel and nip off the anthers. Cover the emasculated flower bud with the straw tube by slowly inserting calyx tube into it. Next day early morning between 7 AM and 11 AM pollen is collected from mature yellow anthers of the selected male parent and dusted on the receptive stigma.

Breeding objectives

1. Yield: High yielding bunch groundnut with dormancy suitable for dry land conditions is important target.
2. Quality: Important quality traits includes high shelling percentage (>75%), high oil content (>50%), high sound maturity kernel (SMK) and table purpose varieties in groundnut.
3. Breeding for biotic stress resistance: Among diseases rust, leaf spot and tomato spotted wilt virus, aflotoxin contamination by *Aspergillus flavans* and among pests red hairy caterpillar and leaf miner are causing major damage in groundnut.
4. Breeding for abiotic stress resistance: Drought and cold are the major abiotic stress in groundnut.
5. Early maturity, erect plant type and cultivars suitable for mechanization are other important breeding objectives in groundnut.

Breeding methods

1. Introduction: The varieties developed and released in other states or counties imported and after evaluation may be released for commercial cultivation.
2. Selection: Mass and pureline selection methods are commonly practiced in groundnut.
3. Hybridization and selection: Inter varietal and inter specific breeding methods commonly used to transfer the traits like disease resistance and quality etc.
4. Mutation breeding: Natural mutation or induced mutation can also be used in development of improved varieties.
5. Transgenic breeding: Development of transgenics for resistance to drought, cold as well as for Leaf spot and tomato spotted wilt virus are the major recommendation.

List of some varieties/hybrids released in India

Variety	Pedigree	Character
TMV-9	AH3490 × AH477	Dormancy up to 15 days, Oil- 51.4%
TMV-10	Natural mutant from Argentina	Testa variegated, High oil content 54.4%
Jyothi	Selected from exotic 7	Less incidence of root rot and collar rots, Oil - 53.3%
Vemana (Kadiri 134)	Kadiri 3 × JL 24	Profuse pod bearing without seed dormancy. Tolerant to leaf spots and sucking pests
Jagital 88 (JGL 88-2)	J11 × TGE 1	Spanish bunch type with high shelling percentage and oil content. Early maturing and tolerant to leaf spot.
Narayani (TCGS 29)	JL 24 × Ah 316/S	Early maturing, spanish bunch derivative. Not suitable to high rainfall areas
Kadiri 71-1	Selected from Ah288	Effective in fixing atmospheric nitrogen, Oil - 47%
M 13	Selected from NC 13	Tolerant to leaf spot, Oil - 49%
GAUG 1	AK 10 × AK12-24	Small thin shelled pod having prominent beak, Oil - 50%
TG 1	X-ray mutant from Spanish improved	Bold pods, profused branching but late maturing variety, Oil - 46.5%
Pollachi 2	Pollachi red × AH2105	High shelling %, Oil - 48.6%
GAUG 10	G 224-3 × G 0343	Plant well branched, stem thin, Oil - 50%
MH 1	Selected from Faizpurl-5	Light green leaves , small pods, Oil - 50%
Dh – 3 - 30	Sapnishpraved × US 4	Pod with promenant beak and Drought tolerant, Oil - 46.5%
TMV 11	Selected from east Africa	Big pod with 1-3 seeds thick shelled, Oil - 50.2%
Chandra	Selected from Ah 114	Bold pod kernel, Oil - 47.7%
TMV 12	Pure line selection from Ugandan culture	Medium to bold pod with small constriction, Oil - 51.2%
JL 24	Selection from EC 94943	Early maturing, Oil - 50.7%
Kadiri 2	Nijerian culture MK374	Efficient in fixing the nitrogen, Oil - 49%
Kadiri 3	Selection from Robust 33	Early maturing and suitable for summer culturing, oil - 49%
BG 1	X –ray mutant	Bold pod SMK, (Sound mature kernel) 93-95, Oil - 48%

Co 1	Ah 6279 × TMV 3	High harvest index, suited for summer cultivation, oil 50.4%
BG 1	X ray mutant	Bold pod SMK 93-95, Oil - 48%
BG 2	X – ray mutant	Bold pod SMK 93 – 95, Oil - 49.3%
Kisan	Spanishinpraved×B 31	Small pod with reticulate venation, Oil - 49.9%
M 37	A1 × C6472	Pod with two seeds small beak, Oil - 50.5%
KGR 1	Selection from Argentina	Small seeds suitable for summer cultivation, Oil - 49%
TG 17	Dark green mutant × TG 1	High harvest index and seed dormancy upto 30days, Oil - 49%
M 197	C 501×U 472	Dark green leaves and early maturing, Oil - 51.2%
GG 2	J 11×EC 16659	Early flowering dark green leaves, Oil - 49%
Jawan	J 11× asiriya mutunda	Elongated pod with large beak, Oil - 47.6%
Co 2	EMS mutant from po 1	Medium pod with 1-2 seeds, Oil - 51.4%
UF 70-103	Introduction from USA	Suitable for summer, Oil - 52%
Dh 8	Selection from RS 144	Tolerant to leaf spot and kernel is round at one end and sharp at other, Oil - 46%
Koushal	Selection from T28	Early maturing plant is compact, Oil - 49%
Chitra	Spanish 5B1×EC 1688	Dark green leaves, Oil - 49%
GG 11	M 13×GAUG 10	Big pod than GAUG 10, Oil - 48.6%
ICGS 11	Selection from robust 33-1	Dark green leaves compact plant, Oil - 48.7%
SG 84	Selected from ICGS 1	Plant grow erect with medium dark leaves, Oil - 50%
GG 7		Bold seeds ,high yield recommended for kharif season
GG 15		Higher yielder than M 13
GG 21		Bold kernel size attractive colour
GG 8		High pod yield (15-17%) than JL 24 & TAG 24
GG 16		40-18.6% high yield than M 335 & ICGV 86325
GJG HPS -1		Bold seed variety and high yield than BAU 13, M 335

GJG-9		Tolerant to stem rot and jassids and thrips
GJG-31		Resistance to jassids damage heliothis and bud necrosis
GJG -17		Shelling out turn is 66.2%, Tolerant to stem rot and high yield and oil
GJG -22		Higher pod yield and uniform kernels, Tolerant to collar rot, Shelling out turn is 72.5%
TNAU Co 6	CS 9×ICGS 5	Dark foliage and tolerant to foliar diseases
ALR 3	(R 33-1×ICGV- 1) × (NCAC 17090 × ALR 1)	Suitable for rain fed areas, Resistance to rust and leaf spot
VRI 2	JL 24 × CO 2	Suitable for irrigation cultivation
VRI 3	J 11 × R 33-1	Early maturing variety and suitable for inter cropping
VRI 8	ALR 3 × AK 303	Moderately resistance to leaf spot and rust Shelling 70%
VRIG N 6	ALR 2 × VG 9513	Moderately resistance to late leaf spot and rust and PDBN disease resistance to early drought
VRIG N 7	TMV 1 × JL 24	Moderately resistance to late leaf spot and rust and leaf miner

Economic importance: Ground nut oil is extensively used in cookery as a salad oil and edibles fried in the oil have a pleasant flavor and keep for a comparatively long time. The oil is used in the manufacture of vanaspati. This oil is also used in shaving creams, pomades, cold creames, candles, glycerin and leather dressings furniture creams. Oil cake is used as cattle feed and also applied as manure for paddy, vegetables, bananas and other fruit trees. Groundnut kernels are rich and cheap source of vegetable protein contain Vitamin A, B (Thiamin) and some members of the B_2 group (riboflavin) and Nicotinic acid.

Castor (2n = 20)

Castor bean (*Ricinus communis* L.) is non-edible oil seed crop grown particularly in arid and semi arid regions. India, Brazil, China, Russia and Thailand are the major castor growing countries of the world. Castor seeds contain between 45 to 60% oil that is rich in triglyserides mainly ricinolein. The genus to which castor belongs is *Ricinus* of the family *Euphorbiaceaea*. Center of origin of castor is Ethiopia. The progenitor of *Ricinus communis is*

Wild relatives: Three species are recognized and described, viz. *Ricinus communis, R. macrocarpus* and *R. microcarpus*, the last being found wild in North Africa and Yemen, and as a weed in the Near and Middle East, even in Kazahstan. This is the form that seems to have first entered into cultivation, in Africa. *R. macrocarpus* is a cultigen derived clearly from *R. microcarpus* and *R. communis* exists in both the wild and cultivated state in a very limited area, in tropical East Africa. The castor oil plant was first cultivated in Russia in the middle or the 19th century, in the Caucasus.

Floral Biology: Plant is monoecious, the inflorescence is terminal panicle with male flowers clustered an irregular cymes at the base, female flowers at the top in racemose manner.

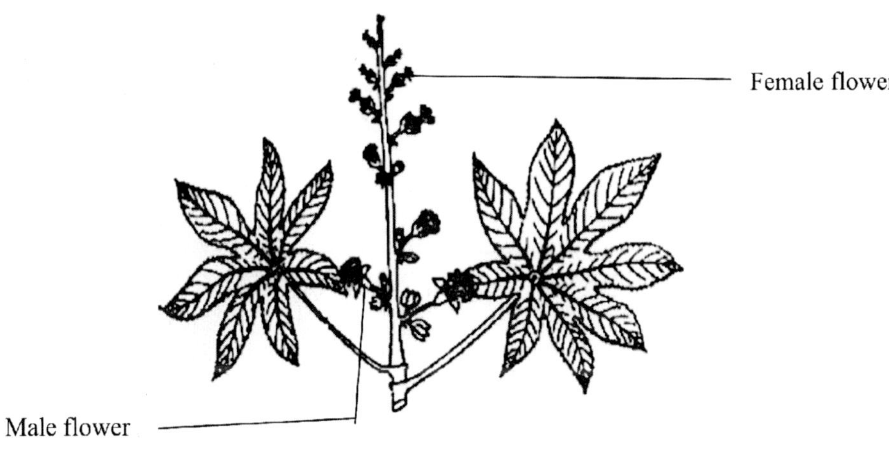

Male flower: Male flower is bracteate, bracteolate, regular, actinomophic short pedicellate, monochlamydeous. Perianth is greenish, in one whorl splitting to 3 to 5 lobes with valvate aestivation. Androecium consists of five stamens which are connate at the base and repeatedly branched. Anthers are dithecous with sub globose divergent lobes.

Female flower: Female flower is bracteate, bracteolate, pedicellate, regular, actinomorphic, incomplete, monochlemydeous and hypogynous. Perianth present in one whorl, five lobed, caduceus valavate aestivation and spathaceous. Gyneocium is tricarpellary, syncarpous, trilocular, superior ovary with single pendulous ovule on axile placentation, styles are three with bifid papillose stigma. Fruit is regma schizocarpic breaking into three cocci. Seed is oblong with white caruncle at micropyle end, testa is crustaceous and mottled, seed contains about 48-56% oil. Endosperm is prominent with membranous cotyledons.

Floral formula:

Male flower: Br, Brl, ⊕, ♂, P_5, A_5, G_o

Female flower: Br, Brl, ⊕, ♀, A_o $G_{(3)}$

Floral diagram:

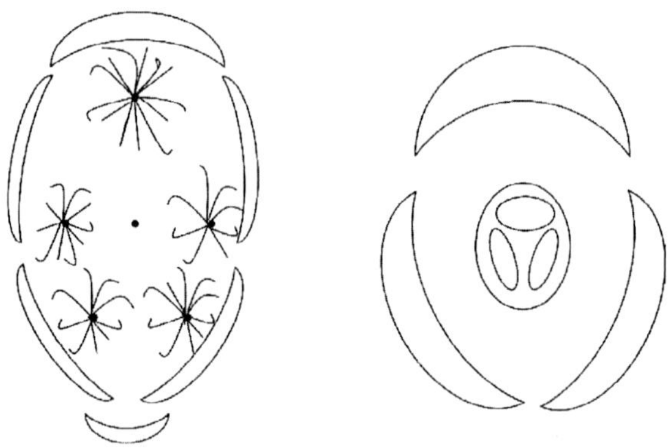

Male flower Female flower

Anthesis and pollination: Cross pollination is the rule in this crop. It is mainly wind pollinated. But insect activity is also seen to some extend since the young leaves just below the inflorescence exude copious nectar at the time of flower opening cause insect pollination. Unisexual flowers, protogynous, elevated position of female flower in the inflorescence, mechanisms to promote wind pollination and nectar glands to attract insect promotes cross pollination. Here male flowers open first (*protoandry*). After one or two days of male flowers opening, female flowers open. However, *protogyny* is also reported. The opening is between 4.30 and 5.00 A.M. Pollen grains are viable for a 2 days and stigma is receptive for 3 days. Each candle takes 10-12 days to complete flowering. To ensue selfing from the selected inflorescence all the male flowers are removed and the female flowers protected with a suitable cover. Artificial cross pollination is done when the stigmas of the retained female flowers become receptive by rubbing the anthers of male flowers collected from the selected male parent.During the rain day old bags are to be replaced with new bags to avoid fungal attack, and free air movement. Emasculation can be achieved by removing or rubbing off the staminate flowers by finger and thumb. For crossing, pollen grains are collected from the desired male parent and are dusted on the stigma of the female parent.

Breeding objectives

1. Yield: Need to develop high yielding short duration varieties/ hybrids suitable for specific situations like rainfed areas, semi-winter conditions, intercropping, mechanical harvesting, saline and poor management conditions.
2. Quality: Hybrids and varieties with medium to bold light coloured seeds with high oil content (>50%) and high ricinoleic acid (>90%) are the consumers' choice. Ricin and (*Ricinus communis* Agglutin) RCA are two highly toxic endosperm proteins present in the deoiled castor cake which makes its cake unsuitable for animal and human consumption as a protein supplement.
3. Breeding for biotic stress resistance: Among the biotic stresses, insect pests (semi looper, castor capsule borer, jassids, white fly and thrips) and diseases (*Fusarium* , *Macrophomina* root rot, reniform nematode and *Botrytis* grey rot) which causes 30–40% of yield losses.
4. Breeding for abiotic stress resistance: Drought under rainfed areas and salinity in major castor cultivation areas are the abiotic stresses which limit castor production.
5. Long duration varieties for dry land areas and breeding for non shattering spineless varieties are the other important breeding objectives.

Breeding methods

1. Introduction
2. Selection: Pureline and mass selections are commonly practiced
3. Hybridization and selection: Inter varietal/inter specific hybridization followed by selection methods are commonly practiced
4. Population improvement: Recurrent section methods are commonly used for population improvement
5. Mutation breeding: Both natural and artificial inducing mutations are used for varietal improvement

6. Heterosis breeding: Breeding programme on development and diversification of stable and superior combining pistillate lines with disease resistance needs acceleration. There is an urgent need to develop CMS lines which is not yet achieved because castor is mono generic and no wild species exist.
7. Transgenic breeding: Development of transgenics for insect resistance and silencing ricin and RCA genes are important consideration.

List of some varieties/hybrids released in India

Varities	Parentage	Special traits
NBCH-2008-2	NRCS 10271	Tolerant to wilt, Botrytis disease.
Jwala (48-1)	E3RC3 11198	
DCH-519	E2RC2 11172	
Deepak (DCH-177)	E1 RC1 11186	
DCH 519	M574 × DCS 78	Resistant to fusarium wilt, leaf hoppers.
Jwala		Resistant to wilt, capsule borer, Tolerant to jassids and botrytis.
DCH 177	DPC 9 × DCS 9	Resistant to fusarium wilt, white fly.
TMV 3		Yield-1100(kg/ht), Oil-45.6%.
DCH 32	LRES 17 × DCS 5	Resistant to jassids and leaf hoppers. Days of maturity-85 to 95.
CH-1		Yield-1800 kg/ha, Oil-49%
J-1		Yield-1200 kg/ha
SA-1		Yield-900 kg/ha
SA-2		Yield-2826 kg/ha
Ginja		Yield-1000 kg/ha
Pb-no.1		Yield-1100 kg/ha,Oil-52%
S-20		Yield-1100 kg/ha,Oil-45.6%
Haritha P.C.S 124	PPL-4 × 48-1	Non lodging, non shattering double bloom, Wilt tolerant
Kiran(P.C.S 136)	Selection from J11-130-1	Red stemmed, double bloomed, Drought tolerant
Jyothi(D.C.S 9)		Wilt tolerant
P.C.H-1		Suitable for rain fed condition. Tolerant to drought.
G.C.H 4		Tolerant to wilt and root rot.

D.C.H 4		Comes to harvest early.
D.C.H 32		Drought tolerant.
Kranthi(P.C.S 4)	(Pb-1 × 157B) × (Ji 44 × 413 A)	Drought tolerant ,Bold seeds.
Aruna	TMV1	Long duration
GC-3		Rsistant to wilt, tolerant to macrophomina root rot
TMV-4	Selection from SA2	
SA 1	TMV 1 × RC 1094	Duration 5 months
TMV 5	SA 2 × S 248/2	Duration 4 months.
LMVCH 1	LRES 70 × TMV	
CO-1	Pure line selection from Annamalai	Bund crop, Fit for raising in vacant areas.
TMV 5	Derivative of SA 2 × S 248/2	Triple bloom, resistant to leaf hopper.
TMV 6	VP 1 × RC 962	Pure, mixed crop, red, double bloom, medium lengthy capsule.
TMVCH 1	LRES 17 ×TMV 5	Triple bloom, semi compact.
YRCH 1	DPC 9 × TMV 5	Triple bloom, spiny, non dehiscent

Economic importance: Castor oil is used in the manufacture of soaps and tooth pastes. Sodium ricinoleate and sulphuricinoleate are used in tooth pastes, due to their strong bactericidal action. Oil is also used as a solid contact lubricant in railway engine and locomotive bearings. Dehydrated castor oil is used in white surface finishes of refrigerators, manufacture of paints and enamels, high quality varnishes, linoleum and oil cloth manufacture.

Castor oil contains glycerides of the unsaturated hydroxyl acid, 90% of a ricinoleic acid imports to the oil certain distinctive properties, high acetyl or hydroxyl value, high specific gravity, refractive index and viscocity, low saponification value, miscibility with ethyl alcohol and near insolubility in petroleum ether. The finest quality of almost colouless castor oil is used in medicinal and aero-engine lubrication.

Sesame or Gingelly (2n = 26)

Sesame or Gingelly (*Sesamum indicum* L.) is one of the oldest cultivated plants in the world. It was a highly prized oil crop of Babylon and Assyria at least 4,000 years ago. Today, India and China are the world's largest producers of sesame, followed by Burma, Sudan, Mexico, Nigeria, Venezuela, Turkey, Uganda and Ethiopia. The genus to which castor belongs is *Sesamum* of the family *Pedaliaceae*. Center of origin of sesame is India and Africa (Ethiopia). The progenitor of *Sesamum indicum is Sesamum angustifolium*

Wild relatives: Only 19 valid species of the genus Sesamum are recognized, their centre being in tropical and subtropical Africa, where the only wild species occur, with the exception of *S. prostratum* in eastern India. The species are described in three sections: I Section, Sesamotypus; *S. heudelotii, S. repens, S. pedalioides, S. mombancense, S. antirhinoides, S. schinzianum, S. angustifolium, S. baumii, S. angolense, S. digitaloices, S. rigidum, S. radiatum, S. dinterii, S. marlothii*, and *S. indicum*; II Section, Sesamopteris; *S. alatum, S. schenckii* and *S. capense*; III Section, Chamaesesame; *S. urostratum*. The majority of the varieties of *S. indicum* ssp. *bicarpellatum* established by Hiltebrandt come from India, those of ssp. *quadricarpellatum* from Japan.

Floral Biology: Inflorescence is solitary, axillary in cyme of 2-3 flowers confined to terminal portion. Flower is pedicellate, bracteate, ebracteolate, zygomorphic with two nectar glands present on either side of the pedicel are supposed to be modified flowers, hypogynous, bisexual, pentamerous, cyclic, di and hetero chlamydeous.

Calyx consists of five green and hairy sepals with gamosepalous, persistent with vavate aestivation. Corolla consists of five petals, gamopetalous, tubular, ventricose, slightly gibbose at base, two lipped, personate, upper lip two lobes usually smaller, the lower bigger and of three lobes, ring of hairs present on the inner side of the corolla towards the base, below the stamens, flowers are white or purple with valvate aestivation. Androecium with four stamens, didynamous, anthers oblong, fifth staminodes maybe present, dithecous, introrse and epipetalous. Gynoecium is bicarpellary, syncarpous tetra locular with many ovules in each locule on axile placentation. Style terminal, stigma forked. There is prominent annular disc present below the ovary. Fruit is oblong and dehiscent loculicidal capsule. Seed is flat, elliptical or obovate, dicotyledons and non endospermic.

Floral formula: Br, eBrl, %, ⚥, $K_{(5)}$, $C_{(5)}$ A_{2+2}, $\underline{G}_{(2)}$

Floral diagram:

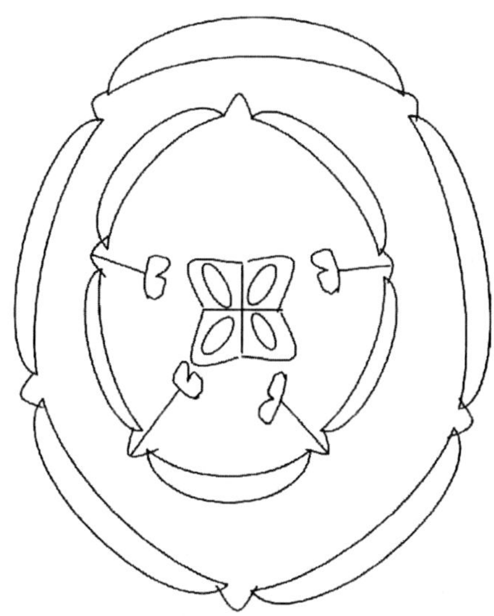

Anthesis and Pollination: Gingelly is a self pollinated (Autogamous) crop. In some varieties cross pollination also takes place to a limited extend of 5-6 per cent. Very high cross pollination between 14 and 65 percent has been recorded in a few varieties in India. Cross pollination may occur due to wind and bee activities. On a bright clear day, the flowers open between 5 and 7 AM. In the mature flower bud, just before the flower opens, the four unripe anthers are much below the stigma which at this stage is not receptive. The anthers begin to burst longitudinally after 4AM in the next day and commence to liberate their pollen. Selfing can be effected by tieing the corolla of the unopened flower which is selected in the previous day evening itself or by smearing a speck of semi-solid clay, on the upper portion of tubular petals of unopened flowers. The emasculation technique in sesame is easy for crossing due to epipetalous nature of the stamens. The flower bud which is expected to open in the next day morning is selected in the previous day evening between 3 P.M. and 6 P.M. and emasculated by just removing the corolla tube in which the stamens are attached. Then, the emasculated flower buds are covered with a piece of soda-straw tube, bent at the top in order to avoid contamination from foreign pollens. During the next day morning, between 7 A.M. and 9.A.M., pollen from the desired male parents were dusted gently on the surface of the stigmas of the emasculated flower buds after removing the soda-straw and again covered. The unemasculated flowers are removed in the female parent. Individual crossed flowers are tagged with coloured thread for the identification of crossed capsules.

Breeding objectives

1. Yield: Varieties with short duration, high yielding, improved plant architecture with adapted crop duration is the important objective
2. Quality: The development of varieties with low or zero anti-nutritional factors like oxalic and phytic acids needs attention for its value addition. In addition, the efforts should also be made to develop low free fatty acid (<2%) varieties.
3. Breeding for biotic stress resistance: The leaf eating caterpillar (*Antigastra catalaulmlis* Dup.) and the gallfly (*Asphondylia sesalili* Felt) are the serious pests of *Sesamum*. Stem and root rot (*Macrophominia phaseoli* Maubl.), phyllody (virus, mycoplasina), bacterial leaf spot (*Pseudomonas sesami* Matkoff) and leaf curl are the important diseases of this crop which needs genetic interventions.
4. Breeding for abiotic stress resistance: Drought is the important problem in this crop.
5. Breeding for non shattering, monostemmed, multi carpellary, white seeded and rice fallow varieties are the other important objectives of this crop.

Breeding methods

1. Introduction
2. Selection: Pure line selection is commonly practiced
3. Hybridization and selection: Inter varietal and inter specific hybridization methods are commonly practiced.
4. Heterosis breeding: Among the various options available for increasing the productivity, heterosis breeding is perhaps the most important way for the vertical yield increase in this crop. China has the distinction of successful exploitation of heterosis in this crop at commercial level with hybrid developed through hand emasculation, GMS/CMS systems and exhibited the yield potential up to 3.0 t/ha. The programme on development of CMS lines through interspecific hybridization needs to be strengthened for exploiting some workable CMS system in this crop for hybrid development programme.
5. Population improvement, polyploidy breeding and embryo rescue technique are the other breeding methods commonly practiced in this crop.

List of some varieties/hybrids released in India

Variety	Parentage	Special trait
TKG-55	TC-25 × TBAU-10	Resistance to major diseases under field condition
JT-22	MT-6 × JLT-3	Resistant to BLB, stem and root rot and leaf roller
AKT-64	N-128 × C-50	Resistance to alternaria, phyllody
AKT-101	N-62-10 × R-19	Resisatance to phyllody, macrophomina root rot and bacterial blight
PKV-NT-11	JLT-9 × NT-4	Resistrant to leaf roller and capsule borer
Guj Til-10		Field resistant to phytopthera and blight disease
TH-6	Punjab Til90 × Line 436602	Single stem
N-8	Selection from local variety	Resistant to major diseases under field condition
Guj Til-1	Selection from local bulk material	Capsules smooth opposite on main stem as well as alternate on secondary branches
CO-1		Erect with profuse branching and short internodes on main stem

TMV-3	Derivative of south local × Malabar	Bushy with profuse branching
TM-4	Pure line selection from Sattur local	Bushy with profuse branching, 50% oil content
TMV-5	Pureline selection from Srivaikundam local	Erect with moderate branching, 51% oil content
TMV-6	Pureline selection from Andhra Pradesh varierty	Erect with moderate branching, 54% oil content
TMV-7	S1 250 × ES 22	Erect, intermediate
VRI 1	Pureline selection from tirukattapalli local	Profuse with moderate branching, 51% oil content
Swetha	E-8 × IS-13	Variety released for cultivation
Chandana (JCS 94)	T-85 × No. 5107	Tolerant to phyllody
Rajeshwari (N 62-39)	Selection from 62-39 of chatpur local	White bold seeded variety
Hima	5039 × AT-1	White bold seeded variety recommended for cultivation
Madhavi	Selection from local A.P	Suitable for sequence cropping
Gouri	Selection from Kokkirapalli, A.P	Tolerant to gall fly
YLM-11		Mature at atime
YLM-17	Vinayak × Kanak	Tolerant to powdery mildew
YLM-66	YLM-17 × P.S 201	Higher yield

Economic importance: Sesamum oil is rich in vitamin E and lacking in vitamin A content. The most useful property of this oil is its high stability and used in curinary purposes. Sesamum seeds are often used in sweet meals. They are roasted and ground into meal and made into cakes used in bakery industry.

Sunflower (2n = 34)

Sunflower (*Helianthus annuus*) widely cultivated for its edible seeds and oil and, to a lesser extent, as an ornamental. The major sunflower-producing countries of the world are the Soviet Union, Argentina, Bulgaria, Pumania, Turkey and South America. Sunflower, as an oilseed crop, was introduced into India in 1969 from USSR with for promising introductions, EC68414, EC68415, and EC 69874. The genus to which sunflower belongs is *Helianthus* of the family *Asteraceae*. Center of origin of sunflower is North America. The progenitor of *Helianthus annuus* is *Helianthus petiolaris*.

Wild relatives: The genus has about 67 species (Heiser, 1978), which includes the annuals and the perennials. These include *H. argophyllus, H. bolanderi, H. dibilis, H. neglectus, H. paradox* and *H. praecox*. The cultivated and wild *H. annuus* have many opportunities for hybridization as they often grow in close proximity in many locations. The wild *H. annuus* is a very common roadside weed in the southern parts of the prairies particularly in Manitoba extending into central United States. A subspecies, *H. annuus* subspecies annuus, which grows in waste places, is most closely related to the cultivated sunflower. *H. petiolaris* and is the main source of male sterile cytoplasm, another annual species that occurs in pockets in Canada, has been known to hybridize with wild *H. annuus*. Several perennial species occur in Canada. The most conspicuous is the *H. maximiliani* which flowers on the roadside in late summer and fall. Some *H. giganteus* occurs in pockets and the *H. tuberosus* (Jerusalem artichoke) is found primarily on riverbanks. This species has been cultivated to a small extent for its tubers.

Floral biology: Inflorescence is a head or capitulum, terminal, heterogamous head, consists of Disc florets (bisexual) in the center. The peripheral florets are ray florets are ligulate unisexual/sterile. The floral parts are enclosed by involucres of green bracts. Floral parts placed on a fleshy receptacle.

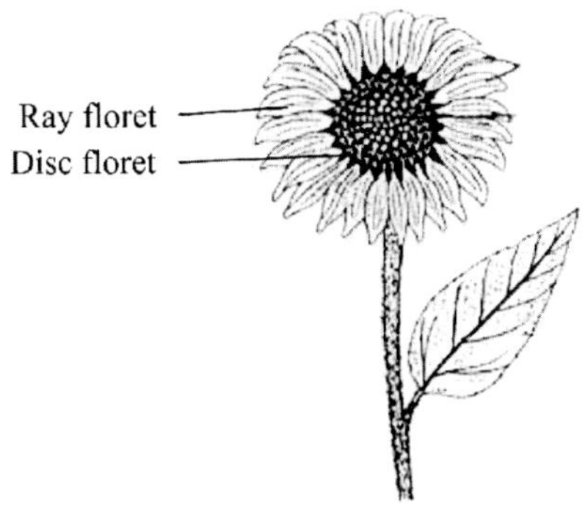

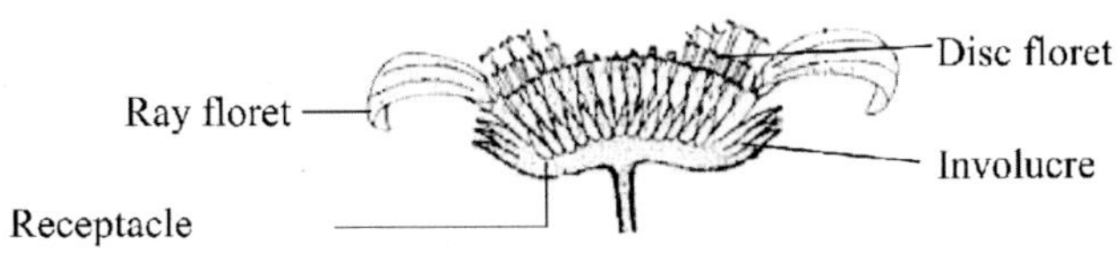

Ray florets are bracteate, ebracteolate, sessile, incomplete cyclic, pentamerous, regularly zygomorphic sterile (neutral) florets, epigamous, dichlamydeous and hetero chlamydeous. Calyx is scaly with 2-3 small scales. Corolla consists of five petals, gamopetalous, ligulate, yellow (large strap shaped) valvate aestivation anterior three lobes are very big and attractive, posterior two lobes are small.

Disc florets are bracteate, ebracteolate, sessile, regularly actinomorphic, bisexual, isomerous, pentamerous, cyclic, di and hetero chlamydeous. Calyx is scaly with 2-3 small scales. Corolla consists of five petals, tubular with valvate type of aestivation. Androecium consists of five stamens epipetalous alternate with the petals. The filaments are free and anthers are fused called syngenesious and each anther is dithecous, basifixed, introrse. Each stamen bears a hood like structure. Gynoecium is bicarpellary, syncarpous, inferior ovary with style passes through ring like syngenesious anthers, stigma bifid and ovary is unilocular with one ovule on basal placentation. Fruit is achene and seed is dicotyledonous, non endospermic, cotyledons contains oil.

Floral formula:

Ray floret: Br, Ebrl, %, K_2, $C_{(5)}$, A_0, G_0

Disc floret: Br, Ebrl, $\oplus$, ⚥, K_2, $C_{(5)}$, A_5, $\underline{G}_{(2)}$

Floral diagram:

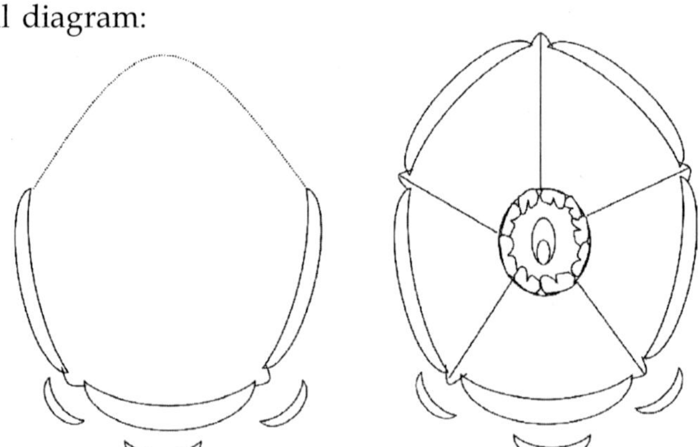

Ray floret Disc floret

Anthesis and Pollination: Sunflower is highly cross pollinated crop mainly through insects (*Entomophily*) and to a limited extent by wind (Anemophily). The flower opening starts from outerside of the head and proceeds towards centre. The head takes 5-10 days for complete blooming depend on size of head and season. Anthesis occurs between 5 to 8 A.M. Pollen viable for 12 hours. Stigma is receptive for 2-3 days. Anthers dehisce early than maturity of stigma (Protoandry), this mainly promotes the cross pollination. For Selfing, the flower head is protected with a suitable cover before the commencement of anthesis in any of the florets and the cover is retained till fertilization is over in all the florets. For crossing, Hand Emasculation is done in the early morning by removing the anthers of the disc florets in 2 to 3 whorls with forceps and the other florets in the head are removed. About 9-10 AM the pollens from desired male parent are collected and dusted on the emasculated head. This process is continued for 2 to 3 days. But in this, head emasculation is difficult. Considering this difficulty, the heads are pollinated without emasculation. On the basis of hybrid vigour, plants are distinguished from the selfed plants. The presence of marker genes for identifying hybrids is also tilized effectively. Male sterility can be induced chemically by spraying of 100ppm GA (Gametocide) during bud-initiation stage consecutively for three days in the morning. Pollination is done in the morning by applying the freshly collected pollen by a small piece of cotton, a hair brush or through fine cloth bag. After each cross, care must be taken to avoid contamination by wiping the hands with alcohol.

Breeding objectives

1. Yield: Development of high yielding short duration varieties suitable for dry land and irrigated conditions is the important target.
2. Quality: sunflower is an important oil seed crop. Oil content ranges from 46-50%. High oil content with monounsaturated (MUFA)/ polyunsaturated (PUFA) mixture of mostly oleic acid (omega-9)-linoleic acid (omega-6) group is important objective.
3. Breeding for biotic stress resistance: Among the diseases, powdery mildew, rust, charcoal rot and *Alternaria* are causing major damage. Among the pests *Heliothis*, grass hopper and jassids causing major problem.
4. Breeding for abiotic stress resistance: Among the abiotic stresses, drought, salinity, heat and cold stresses are important criteria for developing new varieties.
5. Other breeding objectives include development of self fertile lines, Lodging resistance, dwarf plant type and early maturity etc.

Breeding methods

1. Introduction
2. Selection: Mass selection is commonly practiced to develop new variety.
3. Hybridization and selection: Inter varietal and inter specific hybridization methods are commonly practiced in this crop.
4. Recurrent selection: Head to Row and Remnant Seed Method developed by Pustovoit in Russia. By this method oil content is increased.
5. Mutation breeding: Both natural and artificial mutants are used to create mutations.
6. Heterosis breeding: Release of large number of hybrids in the past has broadened the base of ybrids in the country. Still there is need to improve the diversity of the parental lines to achieve the higher level of heterosis in sunfl ower. Diversification of male sterility source may also help in improving stability of the hybrids.

List of some varieties/hybrids released in India

Variety	Parentage	Special traits
DRSF-108	Selection	Yield (kg/ ha)-1200-1400, Days to maturity-95-100, Oil content(%)-37
DRSF-113	Selection	Yield(kg/ ha)-1200-1400, Days to maturity-90-98, Oil content(%)-40
APSH-11	CMS 7-1 A × RHA 271	High yielding, resistant to alternaria, leaf spot and rust. Tolerant to drought. Oil content 40-42%
NDSH-1	CMS 234 A × RHA 859	Non lodging, non branching with slightly drooping habit. Resistant to downy mildew and tolerant to rust, jassids, thrips, head borer and drought.
DRSH-1 (PCSH-243)	ARM 243A × 6D-1	Yield(kg/ ha)-1300-1600, Days to maturity-95-105, Oil content(%)-40

Sunflower varieties and hybrids recommended for different states

State	Variety	Hybrid
Tamilnadu	Co-1,Co-2,Co-3,Co-5	MSFH-1, TCSH-1, TNAU SUF-10, SH416
Gujarat	Gujarat Sunflower, GAU SUF-15	SH-416
Maharashtra	SS-56,Surya, Sidheshwar (LS_11), PKVSF-9, TAS-82	PRVSH-27, MSFH-1, LDMRSH-1, LDMRSH-3, TNAU SUF-6, SH-416
Karnataka	EC 8415	MSFH-1, DSH-1, KBSH-4, RSFH-1, SH-416, KBSH-41, KBSH-42
Andhra Pradesh	-	MSFH-1, APSH-1, SH-416, NDSH-1
Punjab & Haryana	Rumsum record	PSFH-67, PSFH-118
All India	PAC1091, PAC-36, PAC-47, Jwalamukhi (PSCL-5015), Sungene-81, EC 68414, Morden Co-4, DRSF-108	APSH-11, BSH-1, KBSH-1, MLSFH-17, MSFH-8, MSFH-10, TNAU SUF-1, MLSFH-82, DRSH-1, KBSH44.

Economic importance: Sunflower oil is used as edible oil, yellow in colour semi drying oil. Oil is mixed with other drying oils and used for manufacturing of warnishes and paints. Fibre can be extracted from the stem. Fruit can be roasted and after removal of pericarp seed is eaten. Pericarp is used for cattle feed. Sunflower can be used as manure.

Safflower (2n = 24)

Safflower (*Carthamus tinctorius*) is an oilseed crop mainly confined to areas with scanty rainfall. Traditionally, safflower has been grown for centuries from China to the Mediterranean region and all along the Nile valley up to Ethiopia (Weiss, 1971). Presently it is grown commercially in India, the U.S., Mexico, Ethiopia, Kazakhstan, Australia, Argentina, Uzbekistan, China, and the Russian Federation. Pakistan, Spain, Turkey, Canada, Iran, and Israel also grow safflower to a limited extent. It has been grown for centuries in India for the orange-red dye (carthamin) extracted from its brilliantly colored flowers and for its quality oil rich in polyunsaturated fatty acids (linoleic acid, 78%). The genus to which safflower belongs is *Carthamus* of the family *Asteraceae*. Center of origin of safflower is Africa and

Afghanistan. The progenitor of *Carthamus tinctorius* is *Carthamus oxycantha.*

Wild relatives: *Carthamus tinctorius* is not known wild; it seems to be close related to *C. oxyacantha* and *C. lanatus* but crosses between them have never given any forms resembling *C. tinctorius* and it is thought more probable that it arose by mutation.

Floral Biology: Inflorescence is terminal or axillary, homogamous head consists of disc florets surrounded by in volucres of bracts, outer involucres of bracts constricted above the base green spinous and inner involucres of bracts ablong with spiny tips.

Disc florets: These florets are bracteates, ebracteolate, sessile with soft white hairs, actinomorphic, bisexual, tubular, epigynous, yellow or orange in colour. Calyx is modified into pappus. Corolla consists of five petals, ong and tubular with valvate aestivation. Anroecium consists of five stamens, epipetalous, syngenesious, dithecous, basifixed and introrse anthers. Gynoecium is bicarpellary, syncarpous, inferior ovary with single ovule on basal placentation. Fruit is cypsela and seed is dicotyledonous and non endospermic.

Floral formula: Br, eBrl, ⊕, ⚥, K_{pappus}, C_5A_5, $\underline{G}_{(2)}$

Floral diagram: Floral diagram is similar to that of disc florets of sunflower.

Anthesis and pollination: Safflower is basically self-pollinated but bees or other insects are generally necessary for optimum fertilization and maximum yield. Pollination occurs as the style and stigma protrude out of the fused anther tube. Unpollinated stigma remains receptive for several days. Outcrossing in safflower has been reported to range from 0 to as high as 59% in different genotypes in India The outcrossing in genetic male-sterile lines in safflower is reported to be 100%, as no difference in seed yield of male-sterile and male-fertile plants under open pollination was observed. Safflower pollen is transferred by insects and not by wind. The most prevalent pollinator agent in safflower is honey bees. Bees visit the safflower flowers for both pollen and nectar. Each safflower capitulum produces 15 to 60 seeds.

Breeding objectives

1. Yield: Development of high yielding varieties suitable for dry land conditions is important objective.
2. Quality: Due to high proportion of linoleic acid (78%) in its oil, it is considered as healthy oil as it reduces blood cholesterol but makes shelf life of this oil very short and less suitable for frying purpose for its use in food industry. Appropriate reduction in linoleic acid and increase in oleic acid will eliminate this problem and will maintain the tag of healthy oil of this crop. Another aspect of safflower oil quality is increasing the gamma tocopherol content, which is antioxidant in nature.
3. Breeding for biotic stress resistance: Pests like Prodenia and Heliothis are important pests.
4. Breeding for abiotic stress resistance:
5. Varieties with non spiny and thin shell will promote the oil content. So these are also considered as important breeding objectives.

Breeding methods: The breeding methods are similar to that of sunflower.

List of some varieties/hybrids released in India

Variety	Special traits
N-630	20% increase in yield
Manjira (APRR-2)	High yielding with 42.5% oil content.
Sagar mutyalu (APRR-3)	Resistant to rust, suitable for late sowing in rabi rainfed black soils, 35.6% oil content.
A-300	Susceptible to rust

N-62-8	17% increase in yield over N-630
A-1	Suitable under scanty rainfall
Tara	18% increase in seed &oil yield
Bhima	34% increase in yield over Tara
Nira	Specially released for irrigated conditions
Gima	Moderately wilt tolerant
HUS-305	High oil ,salt & wilt tolerant
Sharda	Non spiny
JSI-7	Non-spiny
DSH-129	20-30% higher seed yield than A-1
A-2	High oil
JSI-73	Non-spiny
NARI -6	Non- spiny & High oil, tolerant to foliar and wilt diseases, produces dark red flowers.
Nira (Spiny variety)	Early maturing, recommended for limited irrigations.
NARI-NH-1 (First non-spiny hybrid in the world)	Non-spiny, high oil containing, tolerant to foliar and wilt diseases.
NARI-H-15 (Spiny hybrid)	Early maturing, tolerant to aphids and wilt
NARI-38 (Spiny variety)	Tolerant to wilt
NARI-H-23 First TGMS-based spiny hybrid	Higher tolerance to foliar and wilt diseases than NARI-H-15

Safflower varieties and hybrids recommended for different states

State	**Variety**
Tamilnadu	Co-1**, K-1
Bihar, Orissa, Rajasthan	A-300, A 1, S 144, Sagarmuthyalu
Maharashtra	Bhima (S4), Girnar (JLSF-88)*, N-62-8, Tara, NARI-6** Nira (NRS-209), Sharada (BSF-168-4), Nagpur-7, N-630
Karnataka	A-2, A-300, A-1, S-144
Andhra Pradesh	Manjira (C-438), Sagarmuthyalu (APRR-3)
Madhya Pradesh, Chattisgarh	JSF-1, JSI-7**, JSI-73, JSF-97** , JSF-99,
Uttar Pradesh	Malavika Kusum (HUS-305), T-65**
All India	DSH-129*, MKH-11*, Parbhani Kusuma (PBNS-12), NARI-NH-1 (PH-6)**/*, NARI-H-15*, Phule kusum, NARI-6**

** Non-spriny

Economic importance: The oil is used for culinary and illuminating purposes and for soap manufacture. Charred safflower oil is used for healing sores and in rheumatism. The seeds are used as a diuretic and tonic. Safflower dyes are used for dyeing cotton, wool and silk fabrics. Oil cake is used as cattle feed. Seed husk is used in the manufacture of cellulose, insulation sabrasives etc.

Niger (2n = 30)

Niger (*Guizotia abyssinica*) is an oil crop cultivated in Indian subcontinent and East African countries. Niger seeds contain about 40% edible oil with fatty acid composition of 75–80% linoleic acid, 7–8% palmitic and steric acids and 5–8% oleic acid. The genus to which niger belongs is *Guizotia* of the family *Asteraceae*. Center of origin of niger is Ethiopia. The progenitor of *Guizotia abyssinica* is *Guizotia scabra* subsp. *scabra*.

Wild relatives: The genus *Guizotia* consists of six species, of which five, including niger, are native to the Ethiopian highlands. Those are *G. abyssinica, G. scabra, G. arborescens, G. reptans, G. villosa* and *G. zavattarii*. *Guizotia scabra* contains two subspecies, *scabra* and *schimperi*. *Guizotia scabra* subsp. *schimperi*, known locally as 'mech,' is a common annual weed in Ethiopia.

Floral Biology: Inflorescence is axillary or terminal, heterogamous head and covered by involucres of bracts at the base.

Ray florets: These florets are bracteates, ebracteolate, sessile, incomplete unisexual, female, cyclic, zygomorphic ligulate epigynous, pentamerous, di and heterochlamydeous. Calyx is reduced to a short rim of pappus (scaly). Corolla consists of five petals with gamopetalous, strap shaped, yellow colour with valvate aestivation. Gynoecium is bicarpellary, syncarpous, inferior ovary, style terminal, stigma bifid, unilocular with one ovule on basal placentation.

Disc floret: These florets are bracteates, ebracteolate, complete, epigynous, regular, actinomorphic, bisexual, cyclic, isomerous, pentamerous, di and hetero chlamydeous. Calyx is scaly and pappus. Corolla consists of five petals, gamopetalous, tubular with valvate aestivation. Androecium consists of five stamens, epipetalous, syngenecious, dithecous, basifixed and introrse anthers. Gynoecium is bicarpellary, syncarpous, inferior ovary, style terminal, stigma bifid, unilocular with one ovule on basal placentation. Fruit is cypsela/ achene and seed is dicotyledonous and non endospermic.

Floral formula:

Ray floret: Br, eBrl, %, ♀, K_{pappus} $C_{(5)}$ $G_{(2)}$

Disc floret: Br, eBrl, ⊕, ⚥, K_{pappus} $C_{(5)}$ $A_{(5)}G_{(2)}$

Floral diagram:

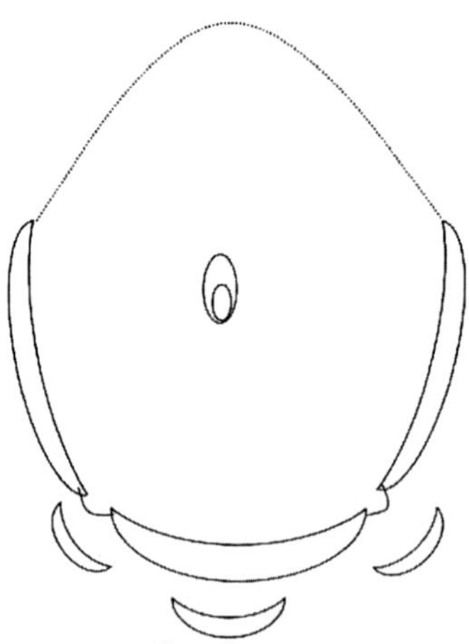

Ray floret

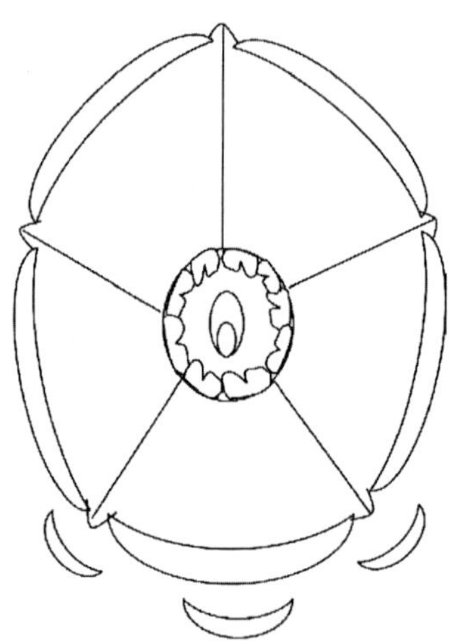

Disc floret

The breeding objectives and methods are similar to that of sunflower.

Safflower varieties and hybrids recommended for different states

State	Variety
Tamilnadu	Paiyur-1, KRN-1
Bihar	Birsa Niger 1, N-5, Ootacamund
Rajasthan/Maharashtra/ NEH/Uttar Pradesh	Sahayadri (IGP-76)
Orissa	GA-10 (Shiva) (Deomali), Bhavani (GA-5), Sahyadri, KRN-1
Andhra Pradesh	KRN-1 (No. 71)
Madhya Pradesh	Birsa Niger-1, N-5, Ootacamund, Jawahar Niger Composite-1
Karnataka	Shrilekha
Gujarat	RCR-317
West Bengal	KRN-1, Sahyadri

Economic importance: The oil is mainly used for culinary and lighting purposes and also used for some pharmaceutical purposes as sesamum oil. The cake contains about 4.5% nitrogen and it is used for manure and as a cattle feed.

Mustard (2n = 16, 18, 20, 22 and 36)

Rapeseed mustard (*Brassica nigra*) is the third most important source of vegetable oil in the world and is grown in more than 50 countries across the globe. China, Canada, India, Germany, France, UK, Australia, Poland and USA are the major cultivators of different species. The genus to which it belongs is *Brassica* of the family *Brassicaceae* or *Cruciferae*. The group of *Brassicas* comprises of six cultivated species, namely, *Brassica campestris / rapa* (2 n = 20, AA), *Brassica nigra* (2 n = 16, BB) and *Brassica oleracea* (2 n = 18, CC) are diploids; *Brassica juncea* (2 n = 36, AABB), *Brassica napus* (2 n = 38, AACC) and *Brassica carinata* (2 n = 34, BBCC) are digenomic tetraploids, which evolved in nature following hybridization between the constituent diploid species.

Floral Biology: Inflorescence is terminal or axillary, corymbose receme (long raceme). Flower is ebracteate, ebracteolate, pedicellate, complete, hypogynous, bisexual, actinomorphic, isomerous tetramerous, cyclic di and hetero chlamydeous.

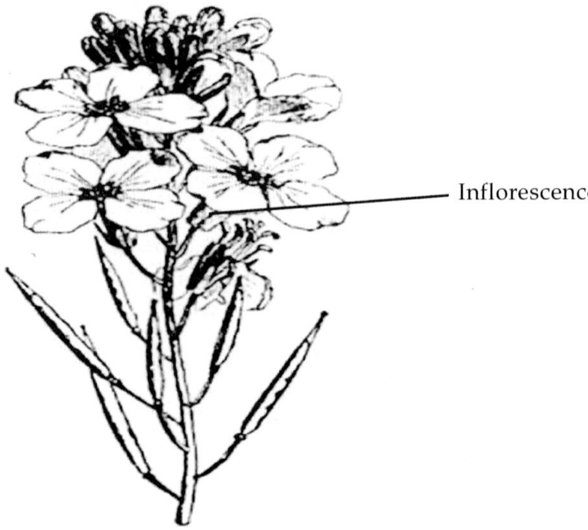

Calyx consists of four sepals, green in colour, poly sepalous with imbricate aestivation. Corolla consists of four petals, polypetalous, cruciform with slenderclaws, yellow in colour with valvate aestivation. Androecium consists of six stamens, 6 (4+2) tetradynamous, the outer two being the shorter, anthers are basifixed, dithecous and introrse anthers. Gynoecium is bicarpellary, syncarpous, superior ovary with false septum called replum is present, number of ovules are present on parietal placentation style is short with capitates stigma. Fruit is siliqua or silicula and the shape of the fruit varies thin long cylindrical or short bulged structure. Seed is dicotyledonous and non endospermic.

Floral formula: eBr, eBrl, ⊕, ⚥, K_5, C_4 A_{2+4}, $\underline{G}_{(2)}$

Floral diagram:

Anthesis and Pollination: *B. nigra* is cross-fertilizing. Pollination is by insects, for example various bee species and pollen beetles such as *Meligethes* spp.

Breeding objectives

1. Yield: Achievement of high yielding genotypes by developing ideal plant type through increasing branch number, pods per plant, seeds per pod and seed size
2. Quality: Quality is an important concern in the times to come. Oil content also needs to be enhanced from the average 38–39% to that of >45%. Breeding of Canola types Indian mustard varieties is the need of the hour to make this crop globally competent.
3. Breeding for biotic stress resistance: *Alternaria* blight and *Sclerotinea* stem rot are the important diseases in this crop. Among pests, aphid and painted bug are important.
4. Breeding for abiotic stress resistance: Drought, high temperature, frost, winter hardiness and salinity are the important targets.
5. Early maturity, herbicide resistance, shattering resistance and meal quality are the other important breeding objectives.

Breeding methods

1. Introduction
2. Selection: Simple selection like mass selection commonly practiced in this crop
3. Hybridization and selection: Inter varietal and inter specific hybridization methods are commonly practiced.
4. Population improvement: Recurrent selection method is followed for population improvement
5. Heterosis breeding: In order to increase the yield potential of *Brassica,* hybrids are one of the most viable options for breaking the yield barriers. Diverse cytoplasmic sources may give high heterotic hybrids under three line hybrid development programme.
6. Using mutations, embryo rescue techniques are the other methods also commonly practiced in this crop.

List of some varieties/hybrids released in India

Variety	Parentage	Special traits
Pusa mustard (PM-30)	-	38% oil content
(CS-58)	-	Salinity tolerant
Pusa double zero mustard 31	-	Duration - 144days, <2% erucic acid.
RH 749	-	Salinity tolerant
NRCHB -101	-	Water stress tolerant variety.
Pusa Agrani	-	High oil content, Water stress tolerant variety.
Pusa Mahak	-	-
NPJ-112	-	-
CS 56	RH 851 × Pusa bold	Salt tolerant, duration 202 days.
CS54	B380 × NDR 8603	Bold seeded salt tolerant, duration160 days.
CS 52	Selection from DIRA 343	Salt tolerant, medium grain, 135 days.
Pusa jaikisan (BIO -902)	Somaclonal variety of varuna	Maturity 110 -115 days, bold seeded varieties.
PUSA BOLD	varuna×BIC 1780	Semi compact branching, Duration 140 -150 days.
Kranthi	-	135 days, 40% oil content
Argani	-	High temperature tolerant.
Arawali(RN 393)	-	-

Economic importance: The leaves of young plants of rape and mustard are used as green vegetables. The plants are used as green fodder for cattle. The seed contains 30-48% oil. The chief use of oil in India is for edible purposes. The seed and oil are used as a condiment in the preparation of pickles and for flavouring curries and vegetables. Mustard oil along with rock salt is used as a dental solution in the diseases of the gum. It is also used as massage in many paralytic diseases of the nervous system.

❏❏❏

Chapter-9

Breeding in Fibre Crops: Cotton, Sun Hemp, Jute and Kenaf

Cotton (2n = 26, 52)

Cotton (*Gossypium hirsutum*) is currently the leading plant fibre crop worldwide and is grown in tropical and sub-tropical regions of more than 80 countries of the world. It is mainly distributed in USA, India, China, the Middle East and Australia, where climatic conditions suit the natural growth requirements of cotton. The genus to which it belongs is *Gossypium* of the family Malvaceae. Center of origin of cotton is Central Africa. The progenitor of *Gossypium hirsutum is Gossypium africanum* and *G. raimondii.*

Wild relatives: The 67 species established in the genus by Js. I. Prohanov are not considered acceptable and even the 33 given by Mauer are thought to be reducible to 21, only 5 of them being cultivated, 2 in the old world (Asia) and three in America. According to Bacsely (1940), a possible origin of *Gossypium hirsutum* is from cross between Asiatic cultivated cotton (*Gossypium herbaceum*) and American wild species (*Gossypium thurberi*) followed by chromosome doubling of F_1 in nature. The new world species are distinguished from those of the old world by having leaves broader at the base, less compact wood, upright pedicels, long corolla teeth not incurned, nectaries both within and outside the flower, both without pubescence, and, finally, oval as opposed to round seeds. All species of cotton are perennials, though some are grown as annuals under cultivation. All are found growing wild and all cross to a greater or lesser degree with one another. They occur in every continent except Europe and are found sporadically also on many of the Pacific islands.

1. *G. sturtii* (2n = 26) grows wild in central and southern Australia.
2. *G. robinsonii* grows wild on the north-west shores of Australia (Nickol Bay).

3. *G. aridum* (2n = 26) grows wild in Mexico, in the coastal state of Sinaloa and to the south of Colima.
4. *G. armourianum* (2n = 26) grows wild on the San Marcos island off the eastern shore of Mexican California.
5. *G. harknessii* (2n = 26), regarded by Prohanov as a synonym of *G. trilobum*, is found-wild on the shores of southern California.
6. *G. klotschianum* (2n = 26) is endemic in the Galapagos
7. Islands, though Var. davidsonii grows on the Revilla Gigedo islands and in the southern part of the California peninsula in Mexico.
8. *G. raimondii* (2n = 26) grows in the Cayamarca and libertad provinces of Peru.
9. *G. thurberi* (2n = 26) grows in Arizona and in the Mexican states of Sonora and Chihuahua.
10. *G. trilobum* grows in central and southern Mexico.
11. *G. triphyllum* grows in south- west Africa, from southern Angola to Damaraland.
12. *G. anomalum* (2n = 26) grows in south-western Africa, Angola, west equatorial Africa, the Sudan, Cordofan, Eritrea and Somali land.
13. *G. areysianum* grows in southern Arabia and has been found on rocks 100km. from Aden. It is extremely drought resistant.
14. *G. stocksii* (2n = 26) grows in Pakistan, south-eastern Arabia and Somaliland.
15. *G. somalense* grows in Somaliland, Kenya, the Sudan and in the north of French Equatorial Africa.
16. *G. arboreum* (2n = 26) is found wild in Pakistan, on the western boundaries of the Deccan and also apparently in Ceylon. In cultivation it is found throughout the east. Being the oldest cultivated species it is highly polymorphic and a number of geographical varieties have been established, those named being var. indicum, var. bengalense, var. cernuum, var. burmanicum and var. nanking.
17. *G. herbaceum* (2n = 26) is found wild in Gujarat, northwestern Pakistan, Baluchistan and eastern and southern Africa, being grown in cultivation in India and Pakistan, Central Asia (in the past), Iran, northern Afghanistan, western China, Asia Minor, the Mediterranean, east Africa and Arabia; var. africanum grows wild

in southern and eastern Africa, in a narrow strip along the Indian ocean and in the coastal areas of Arabia; var. acerifolium grows wild in Gujarat and north-western Pakistan. This was the main species cultivated in Central Asia before the 20th century but it has now been replaced by *G. hirsutum.*

18. *G. tomentosum* (2n = 52) is an endemic species in the Rawai islands.
19. *G. hirsutum* (2n = 52) grows wild in southern Mexico; var. punctatum is found Wild in Haiti, Cuba and Puerto Rico; many semicultivated forms, escapes from cultivation, also exist. It is also described for the island of Tahiti.
20. *G. hirsutum* (not var. punctatum) has become the main cotton in Soviet Russia, where ultra early forms with low-growing sympodia have recently been produced.
21. *G. tricuspidatum* (2n = 52) is found wild in the West Indies and is cultivated only in the tropics of South America and the West Indies; the Bourbon cotton previously classed as *G. purpurascens* belongs here; var. marie-galantae is grown in the West Indies, Ecuador and northern Brazil.
22. *G. barbadense* (2n = 52) comes from Peru, where it is still found wild and as escapes. It has given rise to the Peruvian and Brazilian cultivars.

Floral Biology: Inflorescence is extra axillary terminal solitary born on sympodial branches. Flower is bracteate, bracteolate, bracteoles are three usually foliar, persistent, toothed, palmately lobed bisexual, hypogynous, actinomorphic, on the inner surface of the bracteole, a small gland is present di and hetero chlamydeous. Calyx consists of five sepals forming a truncate cup shortly five toothed with persistant nature with valvate aestivation. Corolla consists of five petals, polypetalous, contorted with twisted aestivation. Androecium consists of numerous stamens, monodelphous stamina column unites with base of petals, anthers are monothecous, reniform or kidney shaped. Gynoecium is tricarpellary, syncarpous, superior ovary, penta locular with many ovules on axile placentation, style is terminal, passing through the staminal tube, stigma blunt or flattend. Fruit is loculicidal casule called as boll, the contents of each individual locule termed as lock. The locules or parts of pericarp collectively called as vue.the aborted ovules that are often found in the capsule are called as motes. The long hairs which are thick and white are called lint. The very short and white or coloured hairs are called as fuzz. The lint is the cotton fibre of consumers.

Floral formula: Br, Brl, ⊕, ⚥, $K_{(5)}$ $C_{(5)}A_{\alpha}\underline{G}_{(3)}$

Floral diagram:

Anthesis and pollination: Cotton is an example for often cross pollinated crop. American cotton flowers opens earlier than Asian cottons. Asian cotton opens in between 8A.M. to 10 A.M. temperature mainly affects flower opening. The pollen may directly used on receptive stigma or may be carried out by insects. For selfing, flowers which are expected to open next day morning are bagged with butter paper bags to prevent cross pollination. Tips of buds are also to be tied with thread/smeared with gum of clay mud pellets. Emasculation is mainly done by two prominent methods. Circular cut method, giving a shallow cut at the base of corolla without inuring the ovary, automatically removes stamens along with corolla and stamina tube. For crossing, the opened flower of male parent is brought to stigma of emasculated flower and rubs on the stigma. In straw pipe insertion, a soda straw of 1.5" inserted through stigma over stamina column by rotating in both directions. Sharp edge of straw tube will clip off stamens and they are tube blown off and other end to be bent to avoid contamination by foreign pollen. For crossing 2-3 anthers from desired male parent are put into straw tube to facilitate pollination.

Breeding objectives

1. Yield: Development of early maturing and high yielding varieties is the important target. In cotton high yielding is mainly determined by short stature, boll number, boll size, high lint percentage, compact and sympodial plant habit etc.

2. Quality: Fibre quality is mainly based on lint percentage. Lint % = Weight of fibre/ weight of cotton. From the point of nutrition, a decrease in the content of gossypol in seeds was considered useful for improving the quality of cotton meal. And also there is a need improve the oil quality by reducing the unsaturated fatty acid content of cotton seed oil and to increase the mono unsaturated fatty acid and poly unsaturated fatty acid contents.
3. Breeding for resistance to biotic stresses: American boll worm, pink boll worm, spotted boll worm, whitefly and jassids are major problematic pests in cotton mainly shows negative impact on yield. Besides these, cotton crop also suffers from bacterial blight and cotton leaf curl virus.
4. Breeding for resistance to abiotic stresses: Drought, heat and salt are the main problems.
5. Breeding for staple length, fiber strength, fineness of fiber and increased lint to seed ratio, good picking quality and glandless cotton are the other important breeding objectives.

Breeding methods

1. Introduction
2. Selection: Pure line selection is practiced for maintainers of genetic purity of existing varieties and development of new by selection with in hybrid population.
3. Hybridization and selection: Pedigree method is commonly practiced.
4. Heterosis breeding: India was the first country in the world to deploy hybrids in cotton. Hybrid seed of cotton mainly produced by manual emasculation and hand pollination between good combiners.
5. Transgenic breeding: Resistance to insects through transgenic technologies was identified as important achievement in the cotton breeding. E.g. BtCotton. And also use of this approach for hybrid seed production was considered as important attainable goal for improving productivity of cotton.

List of some varieties/hybrids released in India

Variety	Parentage	Specific traits
Cambodia cotton	South India	
MCU 1	South India	
MCU 5 VT	*Gossypium hirsutum*	Verticillium wilt resistant
Mahalakshmi (B01 DD)	Ha 11 × (B 134 × CW)	For black soils under rainfed situations
Narasimha (NA 1325)	LRA 5166 × NA 1192	Rainfed medium staple cotton. Spins 30-40 counts. Tolerant to water stress and jassids.
Raghavendra (NDL 5865)	BD 8 (1027 ALF × 2919) /IRAN × 1027 × 2919	It is bushy plant with hairy stem. Drought tolerant.
Lam 389	AV 3373-4 × A 101-1-28	Suitable for rainfed, resistant to bacterial blight and moderately susceptible to alternaria.
CO 2	*Gossypium hirsutum*	Verticillium wilt resistant
LSS	*Gossypium hirsutum*	Verticillium wilt resistant
Cocanada 2	*Gossypium arboreum*	Resistant to bacterial blight
Gaorani 22	*Gossypium arboreum*	Resistant to bacterial blight
Gaorani 46	*Gossypium arboreum*	Resistant to bacterial blight
Lohit	*Gossypium arboreum*	Resistant to bacterial blight
Western 1	*Gossypium herbaceum*	
Selection 69	*Gossypium herbaceum*	
Sujatha	*Gossypium barbadense*	
F 14	*Gossypium hirsutum*	
Bikaneri Narma	*Gossypium hirsutum*	
H777	*Gossypium hirsutum*	
Pramukh	*Gossypium hirsutum*	
SRT 1	*Gossypium hirsutum*	
Narmada	*Gossypium hirsutum*	
L 147	*Gossypium hirsutum*	
PRS 72	*Gossypium hirsutum*	
Mysore Vijay	*Gossypium hirsutum*	
G 27	*Gossypium arboreum*	
HD 11	*Gossypium arboreum*	
LD 133	*Gossypium arboreum*	
Gaorani 6	*Gossypium arboreum*	

AK 277	*Gossypium arboreum*	
Caconada white	*Gossypium arboreum*	Long staple varieties
Saraswathi	*Gossypium arboreum*	
J 34	*Gossypium hirsutum*	
J 205	*Gossypium hirsutum*	
LH 372	*Gossypium hirsutum*	
Vikas	*Gossypium hirsutum*	
Khandwa	*Gossypium hirsutum*	
Rajath	*Gossypium hirsutum*	
Sahana	*Gossypium hirsutum*	
MCU 8	*Gossypium hirsutum*	
MCU 9	*Gossypium hirsutum*	
MCU 11	*Gossypium hirsutum*	
Shyamili	*Gossypium arboreum*	
Maljari	*Gossypium arboreum*	
Srisailam	*Gossypium arboresum*	
Jyothi	*Gossypium arboreum*	
K 7	*Gossypium arboreum*	
Mahanandi	*Gossypium arboreum*	
K 8	*Gossypium arbortum*	
K9	*Gossypium arbortum*	
K10	*Gossypium arboreum*	
DS 5	*Gossypium arboreum*	
HD 107	*Gossypium arboreum*	
LD 491	*Gossypium arboreum*	
Jaydhar	*Gossypium herbaceum*	
Sujay	*Gossypium herbaceum*	
Raichur 51	*Gossypium herbaceum*	
Jayawanth	*Gossypium herbaceum*	
Kalyana	*Gossypium herbaceum*	
Suyodhar	*Gossypium herbaceum*	
Vijalpa	*Gossypium herbaceum*	
G.cot 17	*Gossypium herbaceum*	
Suvin	*Gossypium barbadense*	
Heterosis breeding (Intra -hirsutum)		
H 4	G 67 × American nectariless	

JKh y 1	Khandwa 2(MB) × Reba B50	
Lam cotton hybrid-5 NSP-HH-5	L-104 × L 232	Tolerant to jassids and resistant to bacterial blight and cercospora leaf spot
H 6	Vishnu × SRT 1	
PKVH 2	AK 32 × DHY 286-1	Wider Adaptability
NHH 444	Bikareni Narma×Ac 738	
Savitha	T 7 × M 12	
H 8	G.cot 10 × Surat dwarf	
MECH 4	C 601 × C 219	
CICRHH 1	C 15/2 × B.N	
Fateh	LH 660 × Suman	
Dhan laxmi	H 777 × 1695-175 J	
Maruvikas	SCRF 1 × SCRH 1	
Omshankar	SH 2379 × K34007	
PkV Hy 3	CAk 32 × DHy 286-1R	
PkV Hy 4	CMS based	
DHH 11	CPD 429 × CPD 420	
Inter specific hybrids		
Varalaxmi	Laxmi × SB 289 E	
DCH 32	DS 28 × SB 425 YF	
NHB 12	NS 15 × SB 289 E	
HB 224	LRA 5166 × P 4	
TCHB 213	TCH 1218 × TCB 209	
DHB 105	CPD 428 × B82	
Diploid hybrids		
DH 7	Sujay × G27	
DH 9	4011 × 824	
DDH 2	SM 88 × A 82-1-1	
Mutation breeding		
Indore 2	X-Ray irradiation of Malwa upland 4	High ginning outturn, Spinning performance and jassid resistant
MCU 7	X-Ray irradiation of L.1143EE	Early maturity, increased spinning performance
Rashmi	Gamma-irradiation of MCU 5	Superior quality than MCU 5

Pusa ageti	Gamma-irradiation of Stoneville 213	High ginning outturn, tolerance to jassids
DB.3-12	Spontaneous mutant of Western1	Earliness 20 days, high yield 15-20%
MCU 10	Gamma-irradiation of MCU 4	Suitable under rain fed conditions
DS 1	Gamma-irradiation of G 27	Jassid resistance

Varieties of Bt cotton

Variety	Specific traits
BXN 47	Early maturing variety
BXN 57	Superior ginning outturn
DP20B	Early maturing with Boll gene
DP50B	Wide adaptability, Good storm resistance
DP90B	Full maturing Boll guard variety
DP90RR	Roundup ready gene
DP5690 RR	Excellent strength
DP5415RR	Mid maturing Roundup ready
PM1215	Superior Storm resistance
PM1220	Superior ginning outturn, Wilt resistance

Economic importance: Cotton is a major cash crop which gives three important products fibre, food and feed. Lint fibre is used for clothing, household and industrial articles. Articles of clothing include shirtings, gloves, hosiery etc. house hold articles include bed sheets, covers, towels etc. industrial articles include bags, belting, awnings, tents tarpaulins, insulation, cellulose, plastics etc. cotton seed oil is used in the manufacture of vanaspati, cotton seed cake after extraction of oil is a good organic manure.

Sun hemp (2n = 16)

Sun hemp (*Crotaloria juncea*) is also known as brown hemp, Indian hemp and Madras hemp. It is mainly grown in India, Brazil, and West Pakistan for its fiber. The genus to which it belongs is *Crotaloria* of the family Legumioceae. Center of origin of Sun hemp is India. The genus *Crotalaria* has about 600 species distributed mostly in tropical and subtropical areas of the southern hemisphere and Africa. It has multipurpose use as green manure, fibre and animal fodder crop and has potential as an ornamental plant. Sun hemp is an important source of fibre for the production of paper and ropes in the Indo-Pakistan subcontinent. *C. spectabilis, C. paulina* and *C. mucronata* may have much biomass production as *C. juncea*. However, *C. spectabilis* is toxic. Another species that has potential use in Africa is *C. achroleuca* known as slender leaf rattlebox.

Floral biology: Inflorescence is terminal raceme. Flower is bracteates, bracteolate, pedicillate, complete zygomorphic, cyclic di and heteromorphic, epigynous bright yellow colour. Calyx consists of five sepals, gamesepalous, deeply cleft into two lips the upper two lobed and lower with three lobes. Corolla consists of five petals, polypetalous, papilionaceous, standard broad, wings oblong, keel much pointed, lightly twisted at the apex, descendingly imbricate yellow colour. Androecium consists of ten stamens, monodelphous, dimorphic with five short versatile anthers with slender filaments alternating with five short basifixed anthers with flattend filaments, dithecous and introrse anthers. Gynoecium consists of monocarpellary superior ovary with marginal placentation, ovules few, style long bearded at the top, stigma small and oblique. Fruit is legume; seed is kidney shaped and ex-endospermic.

Floral formula: Br, Brl, %, ⚥, $K_{(2+3)}$, C_{1+2+2} A_{5+5}, $\underline{G}_{(1)}$

Floral diagram:

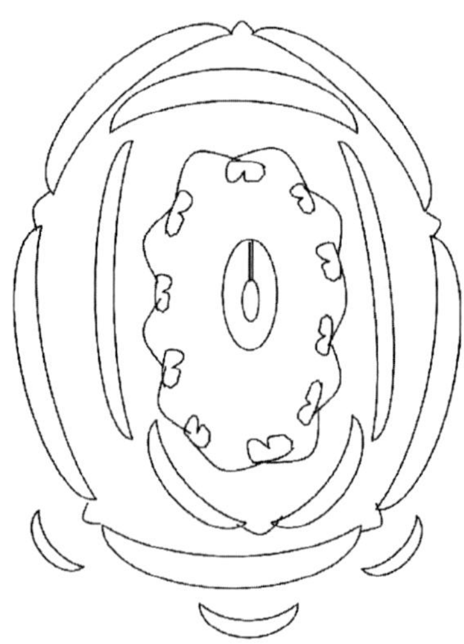

Anthesis and pollination: The lowest flowers of the inflorescence open first and remain open for 2 days. The flowers open between 2

and 3 am. The dehiscence of anthers is complete in between 12 noon to 1 pm in about one hour and a half. The stigma remains receptive on the day of the opening of flower. In the bud the stamens with the linear anthers are longest and dehisce with bud. The filaments of stamens with ovate anthers then elongate, pushing the liberated pollen to the orifice of the keel. The stigma lies at the orifice above the pollen mass. The wings and the keel are articulated by a ball and socket joint. When large bees such as Xylocopa spp. and Vegactile spp. alight on the wings, they catalyse the ball-and-piston mechanism that forces the stigma with a mass of pollen against the abdomen of the insect. Extensive cross-pollination occurs and self-pollination only takes place after the stigmatic surface is stimulated by the bee or mechanically. Slight scratching of stigma with the help of a camel hair brush before pollination also helps in getting good seed set in this crop. A normal pod contains nearly 10 to 15 seeds. Recently, successful efforts in breeding for self-compatibility have been reported.

Breeding objectives

1. Yield: Development of early maturing high yielding varieties for both fibre and seed is important concern.
2. Quality: The characters like length, strength, fineness, colour and uniformity constitute the quality parameters.
3. Breeding for resistance to biotic stresses: Sun hemp is not affected by too many diseases and pests. A timely action to control diseases and pests help in getting a good crop. Wilt, necrosis and anthracnose are the major diseases and top shoot borer, sun hemp moth, flea beetle and root knot nematode are the important pests.
4. Other objectives include improvement of retting percentage, stem quality for paper making and non shattering habit etc.

Breeding methods: Introduction, selection, hybridization and selection, mutation breeding and use of tissue culture approaches are commonly followed in sun hemp.

List of some varieties/hybrids released in India: BE – 1, Belgoan, Chindwara, Jabalpur, Bellary, D – IX, T – 6, ST – 95, ST-55

Variety	Special trait
K-12 (black)	The seeds of this variety are black in colour, yield upto 10-12 q fibre per hectare in 100-120 days maturity. It possesses fibre of good quality.
K-12 (yellow)	The variety is a selection from the old variety K-12 (black). The fibre yield potentiality of the variety is 15 q/ha. It has shown some resistance to wilt disease. The variety has wider adaptability.
M-18	It is a early maturing type with high yield potential. It is resistant to draught, suitable for light soils with low rainfall areas of M.P
M-35	This is more or less similar to M-18. It shows some resistance to shoot borer.
BE – 1	Both yield and fibre qualities are good.
Belgoan	This is suitable for heavy soil and is grown in M.P.
Chindwara	
Jabalpur	
Bellary	It is a high yielder and most popular in Tamil Nadu.
D – IX	It shows resistance to wilt disease and is popular in Maharastra.
T – 6	Most popular in W.B. as green manure crop.
ST – 95	It shows slight resistance to wilt and shoot-borer.
ST-55	Gives high yield of 14 -15 q/ha. It shows moderate resistance to wilt diseases.

Economic importance: Sun hemp is essentially a cordage fibre and is used in the manfacture of ropes, twins, cords and marine cardage. It also used in the manfacture of sail cloth, canuas, matting and rope soles of shoes and sandals.

Jute (2n = 14)

Jute (*Corchorus sps*) is an important natural fibre crop next to cotton. Jute is grown mainly in the delta formed by the Ganges and Bramhaputra rivers in eastern India and Bangladesh. The genus to which it belongs is *Corchorus* of the family Tiliaceae. The genus *Corchorus* includes about 40 species. Two cultivated species are *C.capsularis* (White jute) and *C.olitorius* (Tossa jute). *C. olitorius* is widely cultivated, and has originated from Africa (primary center of origin) with India or Indo-Myanmar region as its secondary center of origin. This species has been reported from Africa, Asia and Northern Australia. The other cultivated species *C. capsularis* is found in Indo-Myanmar and South China region, but not in Africa and Australia. It has originated from Indo-Myanmar region including South China.

Floral biology: Inflorescence is axillary, flower is pedicellate, actinomorphic, hermaphrodite, hypogynous, complete, pentamerous. Calyx consists of five sepals, sepaloid, polysepalous with valvate aestivation. Corolla consists of five petals, polypetalous with twisted aestivation. Androecium contains indefinite number of stamens with free, dorsifixed, dithecous, introrse anthers. Gynoecium is pentacarpellary, syncarpous, superior ovary, pentalocular, axile placentation, two or more ovules per loculus and style and stigma are simple. Fruit is capsule, sub-globose ridged and not breaked.

Floral formula: Br, Brl, ⊕, ⚥, $K_{(5)}\ C_{(5)}\ A_{\alpha}\ \underline{G}_{(5)}$

Floral diagram:

Anthesis and pollination: *Self pollination* is the rule in this crop in both *Corchorus capsularis* and *Corchorus olitorius*. The natural crossing in *C. capsularis* is higher rate and it is due to the wind pollination and insect visitation. Anthesis starts one to two hours after sunrise in *C. capsularis* and about an hour before sunrise in *C. olitorius*. The stamens usually burst before anthesis. To ensure selfing, the flowers may be protected by covering them with fine mesh muslin bags or a polyethylene lantern. Emasculation is done one day ahead or at the time of opening of the flowers. Normally, the first bud to open in an inflorescence is emasculated and the other buds are removed. The buds selected are opened and the stamens are removed with fine pointed forceps. The emasculated flowers are covered with small butter paper bags to protect them from the dew and rain. Pollination is made following morning upto 9.30 AM in case of *C. olitorius* and upto around 11AM in case of *C. capsularis*. The stigma of the emasculated flower is lightly touched with a ripe anther and the pollen is dusted on the stigma.

Breeding objectives

1. Yield: Development of early maturing, high yielding varieties is important concern. Early varieties are generally low yielders whereas late varieties are high yielders. So to combine high yield with earliness is one of the main objectives. Yield is positively correlated with plant height, basal diameter of stem, fibre-stick ratio.
2. Quality: The quality characters are fibre length, fibre strength, fibre colour, luster and percentage and quality of retting etc.
3. Breeding for resistance to biotic stresses: In pests, stem borer and aphids cause greater damage and in diseases *Macrophomena* is major.
4. Development of high seed yield varieties, non shattering varieties is other important objectives.

Breeding methods: Introduction, selection, hybridization and selection, mutation breeding and use of tissue culture approaches are commonly followed in jute.

List of some varieties/hybrids released in India

A number of varieties has been released in India for both the species. These include 13 *tossa* and 12 *white* jute varieties.

White jute: JRC-321, JRC-212, JRC-7447, JRC-4444, UPC-94, Padma, KTC-1, JRC-698, Bidhan Pat 1, Bidhan Pat 2, Bidhan Pat 3, JRC-80

Tossa jute: JRO-632, JRO-878, JRO-7835, JRO-524, TJ-40, JRO-3690, KOM-62, JRO-66, JRO-8432, JRO-128, S-19 (Subala), Bidhan Rupali *tossa*, JRO-204 (Suren).

Economic importance: Jute is a long, soft, shiny vegetable fiber that can be spun into coarse, strong threads. It is mainly used for making twine, rope, and matting. In combination with sugar, the possibility of using jute to build aeroplane panels has been considered. Jute is in great demand due to its cheapness, softness, length, lustre and uniformity of its fiber. It is called the 'brown paper bag' as it is also used to store rice, wheat, grains, etc. It is also called the 'golden fiber' due to its versatile nature.

Kenaf (2n = 36) and Roselle (2n = 36/72)

Kenaf (*Hibiscus cannabinus*) and Roselle (*Hibiscus sabdariffa*), both the species are important jute supplements and show wide adaptability unlike jute. At present both the species are known as Mesta. In trade and industry, jute and mesta crop together known as raw jute as their uses are almost same. Compared to jute mesta is of inferior in quality in respect of fineness, lusture, and colour. Mesta varieties show poor performance in spinning because the fibre is coarse, stiff, brittle and irregular in cross section mesta alone cannot be spun in jute machines unless it is mixed with jute in some proportion. Besides India, the mesta is grown mainly in Argentina, China, Cuba, Egypt, Hewti, Guatamala, Italy, Iran, Indonesia, Mozambique, North Africa, New Guina, Peru, Spain, South Africa, Southern Part of Zimbabwe,Thailand, U.S.A and Russia The genus *Hibiscus* belongs to the family *Malvaceae*. The centre of origin of *H. cannabinus* is Africa and for *H. sabadariffia* is Asia.

Floral biology: Inflorescence/flower is a solitary, with short pedicels, axillary, bracteates, bracteolate (epicalyx) actinomorphic, pentamerous and hypogynous. Calyx consists of five sepals with valvate aestivation. Sepals broader at base with a raised mid rib, with large gland, margin prickly, persistant and gamosepalous. Corolla consists of five petals, polypetalous with twisted aestivation. Androecium consists of numerous monodelphous stamens, anthers are monothecous, reniform and extrose. Base of filaments united which form the stamina column. Gynoecium with superior ovary, pentacarpellary, syncarpous, penta locular, ovules many with axile placentation, stigma as many as carpels (penta fid). Fruit is capsule, pointed and bristled. Seeds are glabrous, dicotyledonous and non endospermic.

Floral formula: Br, Brl, ⊕, ⚥, $K_{(5)}$, C_5, A_α, $\underline{G}_{(5)}$

Floral Diagram is same as like cotton.

Anthesis and pollination: It is mainly out-breeding plant, but up to 30% self-pollination occurs. Flowers open before dayback and began to close about midday. The pistil is functional when the flower opens, while the stamens are not dehiscing until shortly after sunrise. While the pistil is still functional, the stigmatic lobes hang down, almost touching the unopened anthers. Later in the stigmatic become turgid and soon stand above the anthers which loose their pollen. Cross pollination is mainly affected by such as bees.

Breeding objectives and breeding methods are same as like jute.

List of some varieties/hybrids released in India

Variety	**Parentage**	**Specific character**
HS 4288	*Hibiscus subderifa*	Chlorosis appears only above pH7
HS 7910	*Hibiscus subderifa*	
AMV 1	*Hibiscus subderifa*	
AMV 2	*Hibiscus subderifa*	
AMV 3(Surya)	*Hibiscus subderifa*	It is more resistant to root knot nematode and Anthracnose
Kalinga	*Hibiscus subderifa*	Tolerant to stem rot
Bimili 1	*Hibiscus cannabinus*	
AMC 108	*Hibiscus cannabinus*	
HC 583	*Hibiscus cannabinus*	

Economic importance: Kenaf is one of the allied fibres of jute. The main uses of kenaf fibre have been rope, twine, coarse cloth (similar to that made from jute). Kenaf seeds yield an edible vegetable oil. The kenaf seed oil is also used for cosmetics, industrial lubricants and for biofuel production.

□□□

Chapter-10

Breeding in Vegetable Crops: Tomato, Chilli, Brinjal and Bhendi

Tomato (2n = 24)

Tomato (*Lycopersicon esculentum*) ranking 1st in the world for vegetables, accounts for 14% of world vegetable production and it is mainly distributed in Europe, USA, India, Japan and China. The genus to which tomato belongs is *Lycopersicon* of the family *Solanaceae*. Center of origin of tomato is Peru and Mexico. The progenitor of *Lycopersicon esculentum* is *L. pimpinellifolium*.

Wild relatives: Six species are now recognized in the genus *Lycopersicon*. Subgenus Eulycopersicon contains: *L. esculentum* is not known wild; though in cultivation it is grown as an annual, by nature it is a perennial. In this species, two main botanical varieties are grown: (a) var. *pyriforme,* a cultigen originating in Paraguay, Brazil and Peru. (b) var. *cerasifornle,* a slightly trailing form with bilocular fruits and small round fruits, originating in South and Central America. *L. pimpinellifoli*m, which also has round bilocular fruits, comes from the lowlands of Peru, from Ecuador and from the GaLapogos islands. Subgenus Eriopersicon comprises species with pubescent fruits. *L. peruvianum,* a creeping perennial, grows wild in northern Chile and in peru. *L. cheesmanii* has a number of varieties, all growing wild in the galapogos islands. *L. hirsutum* grows wild in Peru and Ecuador, up to an altitude of 2300 m. *L. glandulosum* is found wild in Peru at altitudes between 1900 and 2800 m.

Floral biology: Inflorescence is usually cymose type. It may be terminal or axillary in position. Flowers are bracteate or ebracteate, ebracteolate, pedicellate, actinomorphic, complete, bisexual, pentamerous and hypogynous.

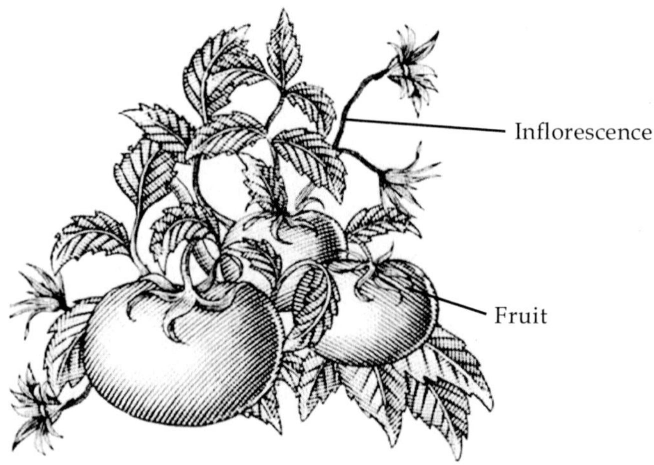

Calyx consists of five sepals, which are gamosepalous with valvate aestivation. Corolla consists of five petals, gamopetalous with valvate or twisted aestivation. Androecium consists of five stamens with epipetalous condition. The anther lobes are large, dithecous, basifixed and introrse. Gynoecium is bicarpellary, bilocular (occasionally unilocular in *capsicum*) and syncarpous with superior ovary. Ovary is oblique in position due to the tilting of posterior carpel to the right and anterior carpel to the left at an angle of 45^0. There are numerous ovules arranged in axile placentation on swollen placeta. The style is terminal and stigma is capitates. And fruit is berry with endospermic and dicotyledonous seeds.

Floral formula: Br, Ebrl, $\oplus$, ⚥, $K_{(5)}$, $C_{(5)}$ A_5, $\underline{G}_{(2)}$

Floral diagram:

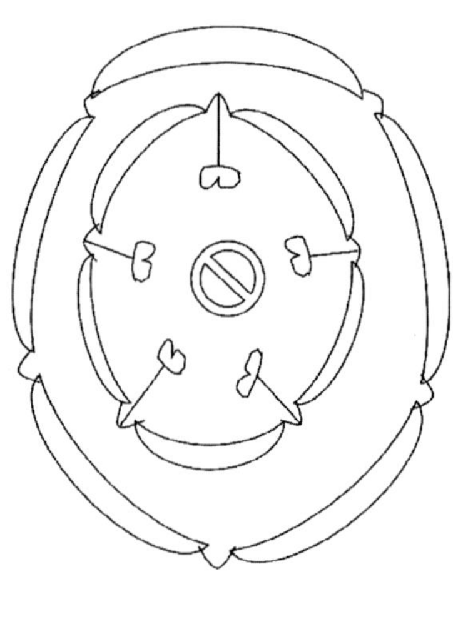

Anthesis and pollination: Emasculation is usually done one day prior to anthesis/flower opening. At this stage, the sepals have started to separate and the anthers and corolla are beginning to change from light to dark yellow. The stigma is fully receptive at this stage allowing for pollination even immediately after emasculation. Anthers are removed as a group with or without the surrounding corolla, by inserting forceps between the sepals to grip the base of the anthers and/or petals which are then removed by a firm but steady pull. Pollen is best applied by slitting the inside of the anthers of mature flowers of the male parent with the forceps in such a way that a small amount of pollen is collected at the tip of the forceps. Protection of pollinated flowers by wrapping with cotton or small pollination bags is essential.

Breeding objectives

1. Yield: Short duration high yielding varieties is the important target.
2. Quality: Fruit quality is mainly depends in feature like large round, uniform size, deep red colour and increased shelf life etc.
3. Breeding for biotic stress resistance: Fusarium wilt, late blight anthracnose, bacterial wilt are the important diseases. Among the pests resistance to fruit borer, whitefly is important consideration.
4. Breeding for abiotic stress resistance: Cold tolerant, drought tolerant, salt tolerant, low temperature tolerant, herbicide tolerant is important abiotic stresses.
5. Development of varieties for prolonged storage, transporation and suitability for processing are the other important breeding objectives.

Breeding methods

1. Introduction
2. Selection: Pure line selection is commonly practiced for development of new varieties
3. Hybridization and selection: Back cross method is commonly utilized in wide crosses or in interspecific gene transfer for resistant to diseases.
4. Transgenic breeding: Breeding for resistance to abiotic stress tomato is sensitive to low and high temp from the stage of germination to ripening. *E.g.,* Transgenic variety "Flavr Savr" was developed for long shelf life and transportation.

List of some varieties or hybrids released in India

Variety	Parentage	Special traits
Azad T 2	HB 5 × Kuber	Plants are determinate, with leaves of curved margin.15mg/100g of Vit C
Arka surabh	Selection from V 685 line	Semi determinate, suitable for fresh market &processing
HS 101	Sel 2 -3 × Exotic cultivar	Early maturing variety with synchronized flowering
Vasundhara	-	Resistant to virus &blight
PKM 1	Induced mutant from local variety called Annaraj	Plants are determinate, suitable for long distance transport
Pusa ruby	Siou × Improved Meerut	In determinate, spreading with few hardy branches
Swarna Baibhav	-	Fruits are deep red, sowing is recommended in Aug – Sep
Kasi Vishesh	*L.hirsutum* BOO13 donar parent	Resistant to tomato leaf curl virus
Pant Bahar	AC238	Plants are semi determinate, more suitable for winter season
Pant T 3	Pure line	Plants are semi determinate, suitable for winter season
Kasha sharad	MTH 6 × Kalyani eunish	Plants are indeterminate, longer shelf life
Vellayani vijai	Introduction and selection from CLN1621 F	High yielding, early maturing variety
Kaliyanpur Angoorlata	Local germplasm collection	Plants are creeping in nature, more keeping quality
HS 110	selection from exotic line	Determinate, large fruits
Swarna lalima	-	Resistant to bacterial wilt
Swarna naveen	-	Have high yield potential
Kasha amrit	Sel 7 × *L.hirsutum*	Determinate variety ,suitable for tomato leaf curl virus prone area
Arka ashish	UC 82 B line from USA	Tolerant to powdery mildew &fruit cracking
Arka ahuti	Selection from Ottawa 60	Suitable for processing
Azad T 6	-	Plants are determinate, fruits round
Bhagyashree	-	Resistant to blight

Solan vijir	-	High yield potential
Arka vikas	American cultivar	Suitable for rainfed &irrigated conditions
Junagarh ruby	Big silari × Pusa ruby	Semi spreading plants
Swarna sampada	-	Resistant to early blight disease
Kasha hemant	Sel 18 × Flora dade	Plants are determinate , red coloured and has yield potential 400 - 425 q/ha
Punjab tropic	From USA	Indeterminate growth habit, fruits are large & round
Mukthi	Selection from LE79	Resistant to bacterial wilt & fruit cracking
Arka abhijit	15SBSB × IIHR 1334	Semi determinate, resistant to bacterial wilt
Arka shreshta	15SBSB × IIHR104	Resistant to bacterial wilt
Arka meghali	Arka vikas × IIHR554	Plants are semi determinate
HS 102	S 12 × Pusa early dwarf	Suitable for both winter & summer
Hisar lalit	HS101 × Resistant Bangalore	Resistant to root rot nematode
Hisar arun	Pusa early dwarf × K1	Early maturing variety
Arka alok	Selection from IIHR 719-1-6	Resistant to bacterial wilt & suitable for fresh marketing
Solan sagun	-	Tolearant to Alternaria
Hisar lalima	Pusa early dwarf × HS 101	High yielding variety
Hisar anmol	Hisar arun × *L.hirsutum*	Resistant to tomato leaf curl virus

Economic importance: The tomato is grown and eaten around the world. It is used in diverse ways, including raw in salads, and processed into ketchup or tomato soup. Unripe green tomatoes can also be breaded and fried, used to make salsa, or pickled. Tomato juice is sold as a drink. Tomatoes, aside from being tasty, are very healthy as they are a good source of vitamins A and C. Cooked tomatoes and tomato products are the best source of lycopene since the lycopene is released from the tomato when cooked. Lycopene is a very powerful antioxidant which can help prevent the development of many forms of cancer.

Chillies (2n = 24)

Chillies (*Capsicum frutescence*) are also called as pungent pepper grown all over the world except in colder climates. Bell peppers are constituents of many foods, add flavour, colour, vitamin C and pungency. It is mainly cultivated in Brazil, Mexico, Spain South and Central America China and India. The genus to which chilli belongs is *Capsicum* of the family *Solanaceae*. Center of origin of chillies is Tropical America.

Wild relatives: The genus *Capsicum*, in which five species are recognized, four of them being grown only in their native country. Subgenus *Campanulatum* comprises one species. *C. bolivianum*, with carnpanulate corolla and small, round, very pungent fruits, grows wild in Bolivia. Subgenus *Rotatum* comprises four species. *C. peruvianum* is an obligate self pollinator and comes from Peru and Bolivia. *C. pubescens* has scented flowers and it comes from Brazil and Colombia and in Europe it is unknown. *C. annuum* is a perennial like all other species but is grown as an annual in cultivation. The speoies comes from Mexico and Guatemala.

Floral biology: Inflorescence is terminal solitary cyme. Flower is bracteates, ebracteolate, pedicillate, pedicel pendent, complete, hypogynous, disc is present, regularly actinomorphic when young and regularly zygomorphic when old, due to the orientation of the ovary in clock wise direction (45^0 angle) the flower becomes zygomorphic, bisexual, cyclic, isomerous, pentamerous, dichlamydeous and heterochlamydeous.

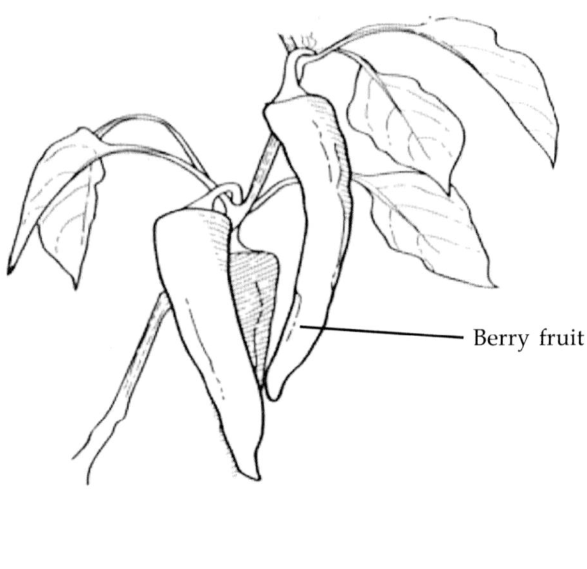

Calyx consists of five numbers of sepals, gamo sepalous, green in colour valvate aestivation and persistant. Corolla consists of five petals, gamopetalous, rotate, white or greenish white in colour, plaited in bud condition with twisted aestivation. Androecium with five stamens, epipetalous, each stamen is having long anther lobes and short filaments, dithecous, basifixed and introrse. Gynoecium is bicarpellary, syncarpous, superior ovary, style terminal, stigma sub capitates, and ovary bilocular with number of ovules on swollen axile placentation. Fruit is a berry, the epicarp and the seeds contain an alkaloid called 'capsasin' pungent. Seed is dicotyledanous, non endospermic, flat, round or reniform.

Floral formula: Br, Ebrl, %, ⚥, $K_{(5)}$, C_5 A_5, $\underline{G}_{(2)}$

Floral diagram: It is similar like tomato.

Anthesis and Pollination: Days required for flowering in chilli crop mainly depends upon the cultivar, temperature, light intensity and duration, soil moisture, fertility status and the age of seedling at transplanting. For crossing, a bud one day prior to anthesis is ideal for emasculation. It can be pollinated on the same day and the bud may be wrapped by thin sheet of cotton which is slightly tightened by giving a side twist. Pollen is transferred to the stigma either from mature undehisced anthers by scooping it out through the lateral sutures with the needle or by touching a freshly dehisced anther to the stigma with the forceps. Percent fruit set in chilli ranges around 20% of the flowers prorduced.

Breeding objectives

1. Yield: High yielding with desirable fruit shape and size is important target.
2. Quality: Superior fruit quality mainly depends on pleasing flavour, high sugar/acid ratio, high pigment content and vitamin C in bell pepper and high capsaicin.
3. Breeding for biotic stress resistance: Fruit rot, Cercospora leaf spot, powdery mildew, bacterial leaf spot, phytophthra root rot, root knot and TMV are the important diseases. Among the pests thrips, mite, aphid, fruit borer are important consideration.
4. Breeding for abiotic stress resistance: Heat, water stress and salinity are the important stresses needs to be concentrated.

Breeding methods

1. Introduction
2. Selection: Pure line selection is commonly practiced
3. Back cross and pedigree methods are applied for selection plants following the hybridization
4. Heterosis breeding: It is one important approach for development of high yielding varieties. The first hybrid in India was Bharat developed by Indo American hybrid seed company, Banglore (1973) followed by several companies
5. Mutation breeding and transgenic breeding are the other breeding strategies commonly followed in this crop.

List of some varieties or hybrids released in India

Variety	Parentage	Special traits
Co.1	Reselection from sattur samba	Fruits have high capsaicin content
Bhgya lakshmi G4 (CA 766)	-	Suitable for green chillies having wide adaptability
LCA 305	LCA 235 × Baruvu mirapa	Tolerant to sucking pests and farly tolerant to virus and leaf spots
Hybrid chilli Co 1	-	Moderately resistant to fruit rot disease
PMK 1	Co 2 × Ramanad mundu	Suitable for rain fed cultivation.
Co 2	Selection from gundu type	Fruits are red in color with high seed content and pungency.
Co 4	Pureline selection from open pollen variety	Suitable for making pickles, low pungency
PLR 1	Kandangadu type	Fruits are medium in size with bulging base and blunt tip
KKM(Ch 1)	-	Early maturity ,suitable for export
KI	Pure line selection from an Assam type B72A	Plants are tall suitable for rainfed cultivation.
Co3	Selection from OP type	Fruits have high oleoresin content (13%)

Economic importance: Dry and green chilli is extensively used as a spice in all types of curries and dishes. It is also used for seasoning of egg, fish and meat preparation sauces, chutneys, pickles, frank fruters, sausages etc. green chillies are rich in vitamin C and A so they serve as good vitamin supplement in addition to their being an appertizer. Simla mirch is rich in Vitamin C. Capsicum preparation is used as counter-irritants in lumbago, neuralagia and rheumatic disorders.

Brinjal/Egg plant (2n = 24)

Eggplant, also known as brinjal or aubergine (*Solanum melongena*), is an important vegetable crop cultivated worldwide, and is the second most important solanaceous fruit crop after tomato. Brinjal is mainly distributed in India, Japan, Indonesia, China, Bulgaria, Italy, France, USA and African countries. Today, Italy, Turkey, Egypt, China and Japan are the leading growers of brinjal. The genus to which brinjal belongs is *Solanum* of the family *Solanaceae.* Center of origin of brinjal is India and Sri Lanka. The progenitor of *Solanum melongena* is *S. insanum*, a species widespread in tropical Asia from Madagascar to the Philippines.

Three distinct cultivated species of eggplants are known: (1) the Gboma eggplant (*S. macrocarpon* L.); (2) the scarlet eggplant (*S. aethiopicum* L.); and (3) the brinjal eggplant itself (*S. melongena*). While the two former are mainly grown in small scale local agricultural systems in Africa, the brinjal eggplant is cultivated worldwide, with a focus in Mediterranean and Asia in particular, where it ranks among the top five most important vegetable crops

Floral biology: Floral biology, floral formula and floral diagram are same like tomato.

Anthesis and Pollination: Flowers generally emerge 40-45 days after transplanting. Anthesis occurs at about 6-8 a. m. in August - September and usually between 9.30 to 11.15 am during winter (December-January). Stigma receptivity is highest during anthesis i.e. flower opening. Anthers usually dehisce 15 to 20 minutes after the flowers have opened. Pollen usually remains viable for a day during summer and 2-3 days in winter under field conditions. For emasculation, the bud is opened gently with the help of fine pointed forceps one or two days before the opening of the bud and all the five anthers are carefully removed. For pollination, freshly dehiscing anthers are picked up and are slit vertically with fine needle to get sufficient pollen at the tip of the needle. Pollen is applied on the stigma of the emasculated flower bud.

Breeding objectives

1. Yield: Development of early maturing, high yielding varieties is the important target.
2. Quality: It mainly depends on rich colour, soft flesh, taste and aroma, and great storage quality. In this nutritive value is mainly depends on high dry matter, sugars, Lower Olanine content, anthocyanin and total phenol contents, low level of polyphenol oxidase activity and orthodihydroxy phenolic compounds to avoid browning of cut fruits
3. Breeding for resistance to biotic stresses: Important diseases are fusarium wilt, verticilium wilt and bacterrial wilt. And pests are eggplant fruit borer, melon aphid, leafhopper and root-knot nematode etc.
4. Breeding for resistance to abiotic stresses: These are drought, low or high temperatures, salinity etc.
5. Upright study plants free from lodging, parthenocary and resistance to herbicides are the other important breeding objectives.

Breeding methods

1. Introduction
2. Selection: Pure line selection is commonly practiced.
3. Hybridization and selection: Pedigree and back cross methods are commonly practiced.
4. Heterosis breeding: It is mainly used for development of F_1 hybrids.

5. Mutation breeding and transgenic approaches (E.g. btBrinjal) are the other methods followed mainly for crop improvement in this crop.

List of some varieties or hybrids released in India

Variety	Special traits
Pusa Purple Long	It is an early maturing and long fruited type variety. Fruits are glossy, light purple in colour, Suitable for spring and autumn plantings, average yield is 27.5 t/ha. It is moderately tolerant to shoot borer and little leaf disease.
Pusa Purple Cluster	It is an early maturing and long fruited type variety. Variety is resistant to little leaf disease under natural conditions.
Pusa Kranti	Good for both spring and autumn plantings under North Indian conditions. Average yield is 14-16 t/ha.
Pusa Barsati	Dwarf and erect growth habit devoid of thorns. Fruits are medium, long and purple with average yield 35.5 t/ha.
Manjri Gota	The fruits are medium-large, round and purple coloured with white stripes. Average yield is 15- 20 t/ha.
Vaishali	Fruits are oval in shape purple in colour with white stripes. The stalks of the fruits bear spines. Average yield is 30 t/ ha.
Arka Navneet	Free from bitter principles with very good cooking qualities. Average yield is 65-70 t/ha.
Arka Sheel	Average yield is 38 t/ha.
Arka Kusmukar	Soft texture with good cooking quality. Average yield is 40 t/ha.
Arka Nidhi (BWR - 12)	Fruits free from bitter principles with slow seed maturity and good cooking quality. Average yield is 48 t/ha.
Arka Keshav (BWR - 21)	Fruits tender, free from bitter principles with seed maturity. Average yield is 45 t/ha.
Arka Neelkanth (BWR - 54)	Fruits tender, free from bitter principles with seed maturity. Average yield is 43 t/ha.
Pusa Ankur	It is an early bearing and becomes ready for first picking 45 days after transplanting. Its fruits do not lose color and tenderness even on delayed pickings.
Bhagamati	Cluster gearing. 2-6 fruits per node, deep purple in colour, drought resistant

Economic importance: Brinjal is widely used in its native India, for example in sambar, dalma, chutney, curry, and achaar. Owing to its versatile nature and wide use in both everyday and festive Indian food, it is often described as the "king of vegetables".

Bhendi/Okra (2n = 72-130)

Bhendi/Okra (*Abelmoschus esculentus)* also termed as lady's finger, mainly cultivated in cultivated in the entire warm temperate and tropical regions of the world for its fibrous fruits or pods containing round, white seeds. It is mainly distributed in Asia, Europe, Africa and United States and Brazil. The genus to which bhendi belongs is *Abelmoschus* of the family Malvaceae. Center of origin of bhendi is India. The progenitor of *Abelmoschus esculentus* is *A. tuberculatus*.

Wild species: The species *esculentus* of genus *Abelmoschus* is a "abeleds" of the tropical and subtropical low altitude regions in Asia, Africa and America, with an extension to the temperate regions of the Mediterranean basin, the species *A. manihot* is cultivated mainly for consumption as a leaf vegetable in Papua New Guinea, the species *A. moschatus* (= *Hibiscus abelmoschus*) is found cultivated and spel-wild in Africa, Asia and America. *A. caillei* is mainly cultivated in humid zone of west and central Africa.

Floral biology: Floral biology, floral formula and floral diagram are similar like cotton.

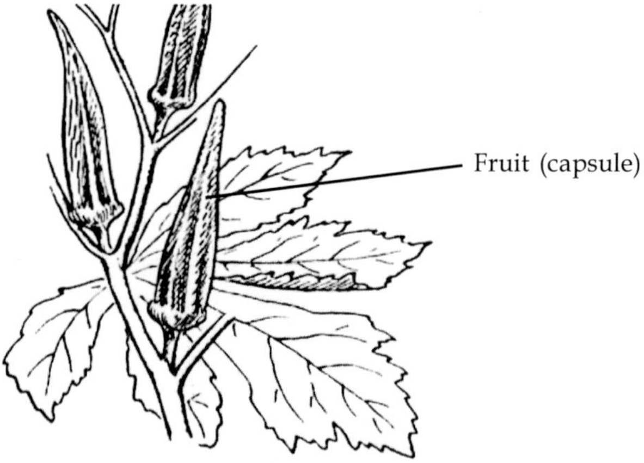

Anthesis and pollination: For emasculation, buds likely to pen the next day are given a slight ring cut at the base of the flower bud with the help of a blade. Petals along with calyx sheath are removed and abeled column and stigma are exposed. The underhisced anthers are removed with a pair of forceps. The emasculated buds are bagged. Next morning, the bags are removed and the emasculated buds are pollinated with the male flowers bagged a day earlier. Pollinated flowers are bagged and abeled. After a few days the bags may be removed. Fruit is elongated, conical capsule.

Breeding objectives

1. Yield: Development of early maturing, late senescing high yielding varieties is the important breeding objective.
2. Quality: It mainly depends on dark green, tender, long, smooth pods and pods free from conspicuous hairs. High yield of seed would be an added advantage.
3. Breeding for resistance to biotic stresses: Varieties resistant to virus diseases such as YVM and leafcurl; fungal diseases such as vascular wilt, Cercospora blight, powdery mildew, fruit rot and damping off and insect pests such as shoot and fruit borer, leafhopper, apdhids, whitefly etc.
4. Breeding for resistance to abiotic stresses: Drought, low or high temperatures and salinity etc. are the important abiotic stresses.
5. Early and prolonged harvest, short plant with more number of nodes, short internodes and pods suitable for processing industry are the other important breeding objectives.

Breeding methods: Introduction, pure line selection, hybridization and selection (pedigree/back cross), mutation and transgenic breeding approaches are commonly followed in this crop.

List of some varieties or hybrids released in India

Variety	Parentage	Specific traits
DVR - 2	PDIB - 1 × PDIB - 3	Resistant to YMV
Janardhana	Induced mutant	Short fruits suitable for canning
PVR - 2	PDIB - 12 × PDIB - 1	Resistant to leaf curl virus
Pusa Kranthi	Abelmoschus esculentus × A.manihot	Resistant to YMV
Pusa sawani	IC1542 × Pusa makhmali	Resistant to fruit borer
Punjab padmini	A.esculentes × A.makhmali	Resistant to YMV
Arka Anamika	A.esculentus × A.manihot sp tetraphyllus	Resistant to YMV
Parbani Kranthi	A.esculentus.sp.pusa sawani × A.manihot	Resistant to YMV

Kashi Kranthi	IC587718 × VRO - 6	Resistant to YMV
Kashi Lila	IC523815 × IIVR -11	Resistant to YMV
Kashi satdhari	IC111542 × IIVR - 20	Resistant to YMV
Kashi Vibhuti	IC 111547 × IIVR -20	Resistant to YMV
Kashi Pragathi	NIC9303 × PK 20	Resistant to YMV
Kashi Bhairav		Resistant to YMV
Sheetal Jyothi		Resistant to YMV
Gjolt -3	Parbhani kranthi × Pusa sawani	Resistant to YMV
GO - 3		Resistant to YMV
G Jolt - 2		Resistant to YMV
GJO -3		Resistant to YMV
Anjitha		Resistant to YMV
Less	Gowri sepatto local × TCR 874	Resistant to YMV
AOH 263		Resistant to YMV
Varsha		Resistant to YMV
Vijay		Resistant to YMV
Vishal		Resistant to YMV
Arka Navneet		
Pusa Makhmali		Resistant to fusarium wilt
Punjabi no 13		Susceptible to mosaic
IHR 20-31		Resistant to YMV
Red Bhendi		Resistant to YMV
Perkins long green		Resistant to YMV
Arka Abhaya		Resistant to YMV
P-7		Resistant to YMV
Red Ghana		Resistant to rhizoctonia solani
Long Green smooth		Resistant to Nematodes

Economic importance: The products of the plant are mucilaginous, resulting in the characteristic "goo" or slime when the seed pods are cooked; the mucilage contains soluble fiber. Okra leaves may be cooked in a similar way to the greens of beets. Okra seeds may be roasted and ground to form a caffeine-free substitute for coffee. Galactomanan in the seeds is used for gum preparation which has several medicinal value.

❑❑❑

Chapter-11

Breeding in Sugar Crops: Sugarcane

Sugarcane (2n = 80)

Sugar cane (*Saccharum officinarum*) is an oldest crop known to man, a major crop of tropical and sub-tropical regions worldwide. It is mainly grown in countries like India, Mexico, Philippines, china, Australia and New Guinea. The genus to which sugar cane belongs is *Saccharum* of the family *Poaceae*. Center of origin of sugar cane is India.

Wild species: Five sugar-containing species are described, *Saccharum officinarum, S. barberi, S. spontaneum, S. sinense* and *S. robustum*. All wild species except *S. sinense* and *S. robustum* are found in India and this is considered to be the area of origin of *S. officinarum* which however is regarded as a hybrid species. *S. spontaneum* is a perennial, polymorphic grass species believed to have originated in India. It is present across southern Asia, east Africa and to the Mediterranean. *Saccharum barberi* originated in northern India and has been exported to other parts of the world. At one time it was the major species cultivated in North America. *S. sinense* is a Chinese sugar cane. *S. robustum* is mainly distributed in New Guinea. Other wild species includes

Saccharum alopecuroidum - Southeastern USA

Saccharum angustifolium - South America

Saccharum arundinaceum - Southeast Asia and New Guinea

Saccharum baldwinii - Southeastern USA

Saccharum bengalense Retz. - India, Pakistan, Iran, Afghanistan

Saccharum giganteum - Southeastern USA, Cuba, Jamaica, Paraguay, Argentina

Saccharum hildebrandtii - Madagascar

Saccharum maximum - Pacific Islands

Saccharum perrieri - Madagascar

Saccharum strictum - from Italy to Iran

Floral biology: Inflorescence is silky panicle called 'Arrow' and may contain as many as 100,000 flowers. The flowers are borne in paired spikelets, one is sessile and other is pedicellate. Both spikelets have two florets. One floret is sterile and represented by a delicate pointed lemma. Other floret is bisexual without lemma. The androecium consists of three stamens with large bilobed anthers that are present in a whorl. The gynoecium comprises of a round ovary on the central surface with a single ovule. Ovary is superior with two styles, stigma, plumose feathery with single ovule on basal placentation. The pistil has two long terminal styles with large brush like stigma.

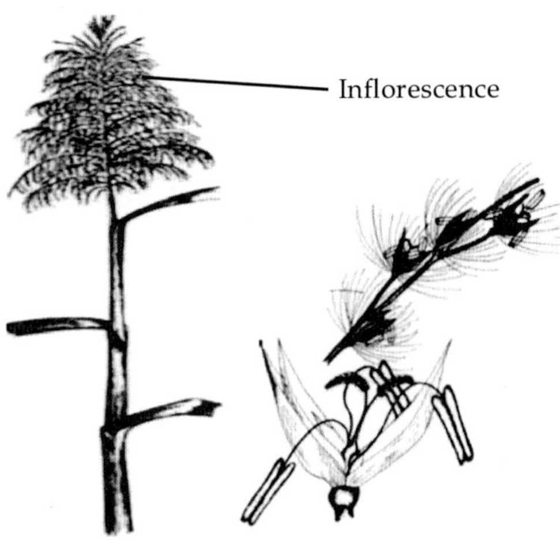

Floral formula: Br, Brl, %, ⚥, P_2, A_3, $\underline{G}_{(1)}$

Floral diagram:

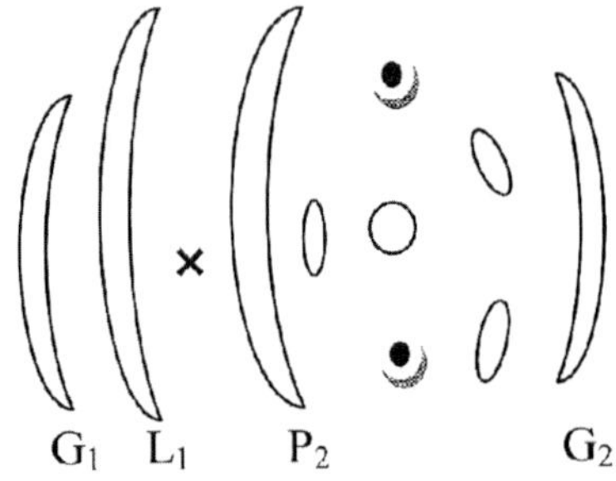

Anthesis and pollination: The flowers open in early morning usually between 5 AM to 6 AM about 7-14 days are required for an arrow to complete flowering. Maximum anthesis between 6 AM and 8 AM. The flowring starts at the top of the arrow and proceeds downwords. Cross pollination normally occurs due to protogynous condition. Selfing is done by covering the arrow with a bamboo frame work or cage which is covered with muslin cloth or polythene paper. Such a cover is commonly called lantern. It prevents accidental cross pollination. This is done preferably during the afternoon hours between 12AM and 4PM. Usually the cover has to be retained in position till the seeds are set. Within one week or 10 days we can get selfed seeds. This selfing method is followed in Sugarcane Breeding Institute, Coimbatore. Due to small size and large number of sugarcane flowers on an arrow, emasculation is not practicable. Crossing done mainly through either with the arrows attached to the parent plants or arrows may be transported to a central crossing area and maintained in a living condition. For pollination, when the arrows of the two parents are enclosed within the same lantern, shaking the arrows occasionally may help to disseminate the pollen. Sometimes arrows from the male parent are collected about 4 AM and brought in to the laboratory, where they are spread on paper and exposed to light at normal temperature.

Breeding objectives

1. Yield: In sugarcane, juiciness of the stem and sugar content are important in yield of sugar per unit of land area. Thick and tall stalks can store juice than thin and short stems. The sugar cane plant stores large quantities of juice containing sucrose in the stalk. So sugar yield is also enhanced by breeding for increased sucrose content.

2. Quality: Important quality parameters of sugarcane include millability, yield of juice, sugar content of juice, and quality of juice. Millability refers to characteristics of the cane that make it possible to recover the sucrose from the stalk by normal methods of extraction. Characteristics desirable for good millability are moderate hardness of rind, good length of fibre, long internodes, and low fibre-sucrose ratio.

3. Early maturity: Cultivars that mature early are desired for milling in the beginning of the season, with progressively later maturing cultivars being milled as the season progresses.

4. Lodging resistance: Resistance to lodging from strong winds and rainstorms require a thick, stout stalk, a healthy and well developed root system, and freedom from diseases or insects injury that weakens the stalk. Tall plants are necessary and desirable for maintaining high yield. So a balance between excessive height and reduced yield from shorter plants may need to be achieved.
5. Resistance to biotic stresses: Rust, smut and red rot mosaic are important diseases. Among pests, sugar cane stem or stalk borer is important problem
6. Resistance to abiotic stresses: Resistance to frost, drought, waterlogging and salinity are important problems.

Breeding methods

1. Introduction
2. Selection: Clonal selection from population of wild clones to isolate genotypes with desirable genes that can be transferred to commercial clones.
3. Hybridization and selection: Hybridization among the commercial clones is followed.
4. Tissue culture techniques: Use of micropropagation has been proposed as means for obtaining pathogen free seed canes for planting or for long term storage.

List of some varieties or hybrids released in India

Variety	Special trait
Vishwamitre	Early maturing, erect, sucrose rich, red rot resistant, suitable for rainfed and water log conditions
Vasudha	Early maturing, suitable for both irrigated and moisture stress conditions resistant to red rot, moderately resistant to smut and tolerant to lodging
Co 8371 (Bhima)	Midlate, resistant to smut, moderately susceptible to red rot, tolerant to drought and water logging
Co 85004 (Prabha)	Early, resistant to smut, moderately susceptible to red rot, good ratooner
Co 86032 (Nayana)	Midlate, resistant to smut, field resistant to red rot, tolerant to drought, good ratooner
Co 86249 (Bhavani)	Midlate, resistant to smut and red rot, good ratooner
Co 87025 (Kalyani)	Midlate, resistant to smut, field resistant to red rot, tolerant to drought and water logging

Co 87044 (Uttara)	Midlate, moderately resistant to smut, moderately susceptible to red rot
Co 87263 (Sarayu)	Early, resistant to smut and red rot, resistant to early shoot borer, tolerant to drought and water logging
Co 87268 (Moti)	Early, resistant to smut and moderately resistant to red rot, tolerant to drought and water logging
Co 91010 (Dhanush)	Midlate, resistant to smut, moderately susceptible to red rot, tolerant to drought
CoM88121 (Krishna)	Midlate, resistant to smut, tolerant to drought, excellent jaggery quality
Co Pant 90223	Midlate, moderately resistant to red rot, resistant to smut, tolerant to drought, water logging and cold, excellent ratooner
CoS 1230 (Raseeli)	Midlate, moderately resistant to red rot, tolerant to top borer, shoot borer and stalk borer
BO 128 (Pramod)	Midlate, moderately resistant to red rot and smut, tolerant to water logging and saline-sodic soil
Co 89029 (Gandak)	Early, moderately resistant to red rot, resistant to top borer and shoot borer, non-flowering, tolerant to drought and water logging.
CoH 2201 (Haryana-92)	Early, moderately resistant to red rot
CoSe92423 (Rajbhog)	Midlate, moderately resistant to red rot
CoSe 95422(Rasbhari)	Early, moderately resistant to red rot
Co 94008 (Shyama)	Early, moderately resistant to red rot, resistant to smut, tolerant to drought and salinity, good quality jiggery
CoS 94270 (Sweta)	Midlate, moderately resistant to red rot, tolerant to abiotic stress, excellent ratooner
CoLK 94184(Birendra)	Early Maturity and good ratooning

Economic importance: Sugarcane is a cash crop gives ready cash to the farmers. It is mainly used for manufacture of jaggery (gul) and sugar. It is used for preparation of juice (raswanti), syrup and also for chewing purpose. Other advantages includes various by products like bagasse, immature tops, molasses, pressmud cake etc obtained from cane are used for different purposes. Immature green tops are used as fodder, bagasse is used as fuel and for making paper, molasses is used for preparing alcohol and other chemicals, pressmud cake is used as manure, trash is used for thatching of huts, mulching and composting and the stubbles are used as fuel or for making compost.

❑❑❑

Chapter-12

Breeding in Narcotics: Tobacco

Tobacco (2n = 48)

Tobacco (*Nicotiana tabacum*) is annually grown herbaceous and tropical crop. It is mainly distributed in the tropical regions of North America, South America, USSR, India and China. Tobacco contains the alkaloid nicotine, which is a stimulant. Dried tobacco leaves are mainly used for smoking in cigarettes, cigars, pipe tobacco. The genus to which it belongs is *Nicotiana* of the family *Solanaceae*. Center of origin of tobacco is tropical and subtropical America.

Wild species: *N. tabacum* is not known wild. The species most closely related to *N. tabacum* are *N. sylvestris, N. glutinosa, N. otophora* and *N. tomentosiformis,* all of which are inhabitants of Peru, Bolivia and north-western Argentina. Artificial hybrids between *N. tomentosa (N. rusbyi)* and *N. sylvestris* closely resemble *N. tabacum* and it is believed that this species has never existed in a wild state but has arisen by hybridization between these species. The artificial hybrids are not identical with cultivated tobacco but indicate the way in which it has probably arisen in the past.

N. rustica is also not known wild. Its cultivation is largely confined to the USSR, where it is grown for producing nicotinic and citric acids, though it is also used for smoking. Its origin is also uncertain, the closest relatives being *N. pavoni, N. undulata* and N. *brachysolen,* all of which grow in Peru and Bolivia.

Other wild species includes

N. quadrivalvis, a wild North American river banks and also in rocky places.

N. multivalvis is also a very early species, growing in sandy places up to 45 N.

N. bigelovii grows in sandy places in North America.

N. torreyana grows in sterile places, sandy and rocks in Central and North America

N. trigonophylla is a herbaceous perennial growing in North and Central America in sandy places and on rocks

N. repanda grows in the plains of Mexico and in Texas.

N. glutinosa grows in stony places in Peru and Bolivia

N. paniculata grows in Peru and Brazil at altitudes of 1000-2000 m.

N. angustifolia grows in Chile. It presents great interest as a source of nicotine.

N. langsdorffii grows in Brazil, Chile, Argentina and Paraguay.

N. sylvestris, a perennial species growing in the mountain forests of Argentina

Floral biology: Inflorescence is terminal in position, stout long peduncle with clusters of cymes arranged in racemose manner. It is a terminal panicle but smaller, well developed inflorescences can develop from axillary buds in the leaf nodes. Flower is bracteate, ebracteolate; usually bracts are present at the base of cymes, complete, bisexual, pentamerous, di and heterochlamydous. Calyx consists of five sepals, gamosepalous, united at back and free at top with valvate aestivation. Corolla consists of five petals, companulate corolla with twisted aestivation. Petals are narrow at top and broader at base. Androecium consists of five stamens with epipetalous condition. The anther lobes are large, dithecous, basifixed and introrse. Gynoecium is bicarpellary, bilocular and syncarpous with superior ovary. Fruit is non endospermic capsule.

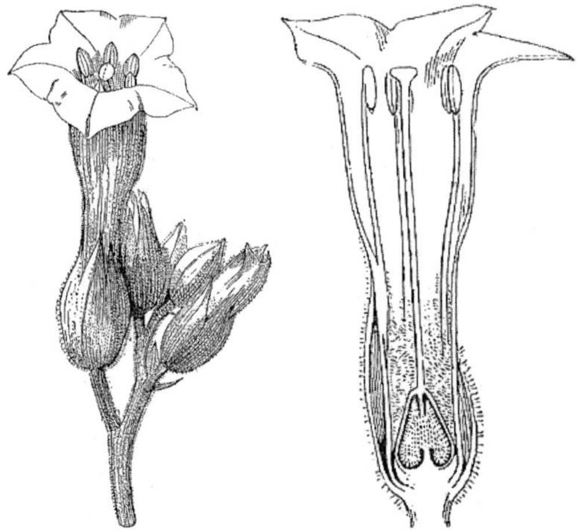

Floral formula: Br, Ebrl, ⊕, ⚥, $K_{(5)}$, $C_{(5)}$ A_5, $G_{(2)}$

Floral diagram:

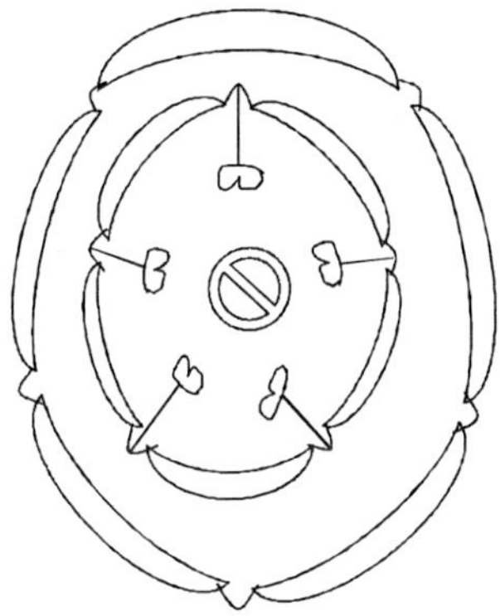

Anthesis and Pollination: Tobacco is a highly self pollinated species and natural cross pollination is usually low. The amount of cross pollination varies with location, season, and activity of pollen vectors. The stigma becomes receptive 24 to 48 hours before anthesis, and pollination of flowers at earlier stages of development is much less successful. Stigma receptivity can be estimated by flower length measurements, and is closely correlated with pigment development in the corolla tip. The period of stigma receptivity after anthesis has not been investigated because dehiscence usually occurs as the flower opens, making such flowers unsuitable for pollination. Flowers selected for emasculation and pollination should be at a stage of development at which anthesis is expected in 12 to 24 hours. For pollination, 24 hours before a pollination is to be made, flowers of the male parent are removed from the plant, placed in a 50ml beaker to which 10 ml of water has been added, and stored in an insect proof room at room temperature. For dusting, a flower from the male parent is selected from the storage beaker and the corolla removed with the forceps, the corolla of the emasculated flower is opened with the index finger by pressing down on the throat on the side opposite the incision. The stigma is thoroughly covered with pollen by brushing with an anther of the male parent. A soda straw is folded at one end and the open end placed over the stigma and style. The corolla is closed around the soda straw and wrapped with a segment of copper wire, thereby protecting the pollinated flower from insect visitations and contamination from other pollen grains.

Breeding objectives

1. Yield and quality: High yield with good leaf quality is important target. The visual quality characters are colour, body, texture, maturity/ ripeness, graininess, hygroscopicity, shatterability, blemish, elasticity, fluffiness, aroma, leaf size, vein colour etc. These characters are subjective based on which the tobacco leaf is graded and purchased by the trader. The manufacturing characters are filling value, equilibrium moisture content, pore volume, elasticity, shatterability, combustibility, lamina-midrib ratio (Strip yield), No. of leaves/kg, lamina weight/unit area etc. Though tobacco leaf contains hundreds of chemical constituents, only few of them will have dominating influence on quality. Nitrogenous and carbohydrate fractions are the two groups of chemical constituents having outstanding effect on smoking quality. Chloride in leaf is also very important as it influences the combustibility and keeping quality of leaf.

2. Resistant to biotic stresses: *Alternaria* leaf spot, blue mold, black shank, broomrape, collar rot, frogeye leaf spot, leaf curl virus, tobacco mosaic, tomato spotted wilt virus and wilt are the important diseases. And among the pests, aphids, hronworm and loopers are major problems.

3. Resistance to abiotic stresses: In tobacco, abiotic stresses such as high and low temperatures; drought, salinity, heavy metals, high light and UV exert a profound effect on the viability, growth, morphology and productivity of plants.

Commonly used breeding methods include introduction, selection, hybridization and selection and use of transgenic technologies for improvement of quality, biotic and abiotic stress resistance traits in tobacco.

List of some varieties released in India

Variety	Special traits
Kanthi	Withstands moisture stress to some extent
Hemadri	-
Siri	-
KST-28 (Sahyadri)	Drought tolerant, mild drought situations
CH-1	Northern Light Soils
N-98	Rainfed conditions of Southern Light Soils
FCH 222	High degree of tolerance to *Fusarium* wilt disease
CH-3	Karnataka Light soils

The above mentioned varieties are belongs to Flue-cured tobacco type only.

Economic importance: Tobacco is a stimulant and the dried leaves of the tobacco plant can be cured and used to produces tobacco cigarettes, cigars and snuff or for pesticide production.

❑❑❑

Chapter-13

Glossary of Taxonomic Terminology

For the identification and description of plants knowledge of technical terms, used in taxonomic literature, is required. In this glossary definitions of taxonomic terms, which are frequently used in the description of plants, are given.

Habitat: Natural home of plant.

Habit: Characteristic form or bodily appearance of organism.

Herb: Any non weedy plant without persistant aerial parts.

Annual: A plant that completes its life cycle, from seed and dies in a single season.

Bi-Annual: A plant that completes its life cycle in two seasons.

Perennial: A plant that continues its growth from year to year.

Ephemeral: A plant with very short life cycle, having several generations in one year.

Shrub: A woody plant of limited height, typically branching at or near the base and not developing much of a trunk.

Tree: A woody plant that produces one main trunk and a more or less distinct and elevated head.

Creeper: A trailing shoots that root throughout most of its length.

Climber: A plant with thin stem with special organs of attachment they cling to neighboring objects for the support.

Lianes: Tropical climbing plants with thin woody stems that twine around the other vegetation.

Epiphyte: A plant growing on other plant using it only for support and not drawing food from it.

Parasite: A plant living in or on another plant or animal host from which it obtains food.

Insectivores: A plant that supplements its nitrogen supply by digesting small insects that becomes trapped on its leaves.

Halophyte: A plant able to tolerate high salt content in the soil.

Aquatic/Hydrophyte: Plants growing in water.

Xerophyte: Plants that grow in deserts or in very dry places, they can with stand a prolonged period of drought.

ROOT

The root is descending axis, growing opposite from the stem, without nodes, mostly developing underground and absorbing moisture from the soil.

Adventitious: Roots arising from any part of the plant body other than the radical.

Tap root: The primary root, forming a direct continuation from the radicle.

MODIFIEDROOTS

Annulated: Root having a series of ring-like swellings.

Aassimilatory: Long slender green hanging root meant to photosynthesis.

Climbing: Roots produced at the nodes by the climbing plant for support.

Fasciculated: When several tuberous roots occur in a cluster or fascicle.

Fusiform: Spindle shaped root, narrowed both ends and a middle part is swollen.

Moniliform: Neckalce – shaped root, like of string of beads.

Napi form: Root considerably swollen at the upper part and sharply tapering at the lower part.

Nodulose: Roots having nodules.

Pneumatophores: Speical roots produced by mangroves for respiration, grows erect, projecting into the air and contain pores at the tip.

Prop/stilt roots: Pillar like roots produced from the stem and branches above the ground that grow obliquely down ward and entering the soil for mechanical support.

Tuberous: thick and fleshy without any particular shape.

STEM

The stem is the part of the main axis of vascular plants, bearing leaves and buds at nodes and reproductive structures.

Acaulescent: Stem less or apparently so.

Areial: Stem which remains above the ground in air.

Arborescent: Main trunk relatively short, plant large almost tree like.

Caudex: the unbranched, erect, cylindrical and stout stem, marked with scars of fallen leaves.

Climbing/scandent: supported by clinging.

Culm: A jointed steam of grosses and sedges with soild nodes and internodes.

Decumbent: Llying on the ground [horizontally] and reproductive organs grow vertically about the soil.

Diffuse: Lossly branching or spreading stem.

Errect: A rigid and strong stem holding itself in an upright position.

Fistular: A stem having a hallaw interior.

Glabrous: Without hairs, smooth.

Herbaceous: Non - woody, like a herb having a green colour and a leafy texture.

Internode: The portion of a stem between two adjacent nodes.

Node: A print of a stem where leaves of branches are attached.

Procumbent: Prostrate, trailing stems, lying flat upon ground.

Scape: A leaf less flowering stem.

Subterranean: Underground steam.

Succulent: Fleshy.

Tailing: A week stem that creeps on the ground without rooting the base.

Virgate: Woodlike, a slender, erect, straight leafy stem bearing only short or no braches.

STEM MODIFICATION

Underground

Bulb: The stem is reduced to disc like structure bear adventitious root on its lower side and bears fleshy, scale-like leaves on the upper side.

Corm: Grows vertically in the soil at a particular depth stores food material, leaves are reduced to scales.

Rhizome: Grows horizontally below the soil.

Tuber: A fleshy underground stem bearing scale like leaves with buds [eyes] in their axils.

In grass plants rhizome does not store food, so the inter nodes and branches remain tubular such stems are called sobale.

Subaerial

Off set: A short, horizontal creeping stem producing a tuft of leaves above and a cluster of below at the tip.

Runner: A slender, weak stems creep on the soil and are rooted at every node. When the internode breaks of, the nodes lead an independent life.

Stolons: In some herbs the long, slender branches which arise from the base of the steam grow obliquely down wards, when these branches touch the soil produce adventitious roots.

Suckers: Underground branches grow obliquely upward from the axillary buds of the nodes present below the soil and directly giving rice to a leafy shoot.

Aerial

Cladode: A branch of a single internode simulating a leaf.

Phylloclade: A flattend branch assuming the form and function of foliage.

Climbing: A week stem ascending by using peculiar structures (Roots lets hooks or tendrils) as support.

Leaf

The leaf is a lateral projection on a stem at a node and subtending a bud, usually expanded and concerned with the manufacture of food.

Aphyllous: Without leaves.

Cauline: Leaves born on the main stem only.

Ramal: Leaves born on the lateral branches.

Radical: Leaves born on the much reduced underground stem.

Deciduous: Falling of soon after maturity.

Persistant: Remaining in place and functioning for one are more years of formation.

Fugacious: Falling almost as soon as formed.

Phyllotaxy: The arrangement of leaves on the steam.

Alternate: One leaf at a node and so arranged that a line drawn on the stem through the leaf based will take a spiral form.

Opposite: two leaves at node on apposite side.

Decussate: A pair of leaves that stands at right angel on the next upper of lower part.

Susperposed: A pair of leaves that stands directly over the lower pair in the same plant.

Turnate: Three leaves at a node.

Whorled: More than two leaves at a node in circle around it.

Distichous: After completing one circle around the stem if the third leaf lies about the first leaf the leaves on stem are seen in two vertical roles.

Tristichous: When the fourth leaf lies exactly about the first one after completing one circle.

Stipulate: Leaf with stipules.

Stipel: A stipule-like appendages at the base of a leaflet of a compound leaf.

Free lateral: Two free stipules born on the two sides of the leaf bases.

Adnate: Two lateral stipules adhering to the petiole upto a certain height.

Inter petiolar: Two stipules lying between the two leaves.

Inter petiolar: Stipules present between the leaf and the axis.

Ochreate: Stipules that form a hollow tube encircling the stem from a node upto a certain height.

Foliaceous: Leafy stipules.

Scaly: Stipules small and dry, born on the two sides of the leaf base.

Tendrillar: Stipules modified into spines.

Petiolate: A stalked leaf.

Pulvinous: Swollen petiole base responsive to vibratious and heat.

Filiform: Long and slender petiole.

Grooved: Petiole provided with long furrow.

Phyllode: Petiole flattend like leaf.

Sessile: Without petiole, the blade being attached directly to the stem.

Decurrent: A sessile leaf with winged leaf base.

Auriculate: A sessile leaf with ear like appendages that partially enclose the stem.

Amplexicaul: Leaf base clasping or embracing the stem completely.

Perfoliate: Basal lobes of a sessile leaf fuse together so that the stem seems to pass through the leaf base.

Lamina: The broad expanded part of leaf.

Simple: A leaf with one blade, with incomplete segmentation.

Compound: A leaf compounded into leaflets.

Leaf let: A single division of a compound leaf.

Petiolule: Stalk of leaf let.

Pinna: Leaf let in pinnate leaf.

Pinnule: Leaf let in pinnately decompounded leaf.

Palmately compound: Leaflets arise from a common base, as fingers of a hand.

Unifoliate: A compound leaf reduced to a single leaf let.

Bifoliate: A compound leaf having the leaflets arranged in two rows along the rachis.

Unipinnate: When the midrib of the pinnately compounded leaf directly bears the leaflets.

Pari pinnate: Uni pinnate with equal number of leaf lets *i.e.* without a terminal one.

Impair pinnate: Unequally pinnate, with a single terminal leaflet.

Bi pinnate: The midrib produces secondary axis which bear the leaf let.

Tri pinnate: The secondary axis produces tertiary axis which bear the leaflets.

Decompound: More than thrice pinnate.

Unijugate: Compound leaf comprises of one pair of leaflets.

Dorsiventral: Flattened and with a distinct upper and lower surface.

Iso bilateral: There is no distinct difference between upper and lower surface internally.

Centric: A leaf which is more or less cylindrical.

Ligule: A small, tongue shaped appendage at the junctions of leaf blade and sheath.

Rachis: The axis of pinnate leaf.

Rachilla: The axis of pinnate leaflet.

Venation: Arrangement of veins in the lamina.

Parallel: When the veins run parallel to each other.

Reticulate: Veinlets are irregularly distributed forming a net work.

Unicostate: With a prominent mid rib.

Multicostate: With a number of prominent veins.

Convergent: Veins which run from the base unite at the apex of the leaf.

Divergent: The leaf palmately lobed and the mid ribs enter into each lobe separately.

Hypoderomous: With the veins sunken in the leaf and not readily visible.

LEAF SHAPE

Acicular: Slender, needle shaped.

Awn shaped: Narrow and gradually tapering to a sharp point.

Cordate: Heart shaped with the notch basal.

Cunnate: Hedge shaped with the broad end apical.

Deltoid: Triangular shape.

Elliptical: Oval in out line, narrowing at two ends and widest at or about the middle.

Falcate: Lamina is sickle shaped or crescent shaped.

Hastate: Arrow shaped with the basal lobes turned outwards.

Lanceolate: Much longer than broad, widening above the base and tapering to the apex.

Linear: Long and narrow, the sides parallel.

Lorate: Strap shaped, ligulate.

Lyrate: The lamina is lobed, the terminal lobe is longer and the lower lobes are smaller.

Obcordate: The lamina is lobed, the terminal lobe is longer and the lower lobes are smaller.

Oblique: The lamina has two unequal halves on either side of the mid rib.

Oblong: Longer than broad, with the margins nearly parallel.

Obovate: The reverse of ovate, broader towards the apex.

Ovate: The lamina is with broad base and a narrow apex.

Reniform: Kidney shaped.

Sagittate: Arrow shaped, triangular with the basal lobes pointing downward.

Auriculate: The base of the lamina is bilobed and partly covers the stem.

Flabelliform: The lamina is lobed and lobes are arranged like a fan.

Rotund: Lamina is circular.

LEAF MARGINS

Entire: The leaf margin is even and smooth.

Undulate/repand/wavy: The leaf margin is wavy having ups and downs.

Dentate: Margins are cut into teeth; teeth are right angles to the margins.

Serrate: Margins are cut into teeth. The teeth are saw like.

Crenate: The leaf margins are round and toothed.

Spinous: Margins provided with spines.

LEAF APEX

Acuminate: Leaf apex is long and slender.

Acute: Sharp with a acute angle.

Obtuse: Apex is rounded or blunt.

Truncate: Apex appears to be cut horizontally.

Retuse: This is an abtuse apex with a slight notch in the middle.

Emarginate: Apex is characterized by a deep notch.

Cuspidate: Apex is like a long, pointed spinous structure.

Murcronate: Having a short, sharp point at the apex.

Cirrhose: Tendril like, having a slender coiled apex.

SURFACE OF THE LEAF

Glabrous: Smooth and free from hairs.

Glacous: Surface covered by wax.

Scabrous: Rough with short and rough hairs.

Pubescent: Covered with short and smooth hairs.

Glandular: Covered with grandular hairs.

Spiny: Covered with spines.

TEXTURE OF THE LEAF

Herbaceous: Leaf is thin and smooth.

Coriaceous: Leaf is thick and leathery.

Succulent: Leaf is thick and fleshy.

Hygroscopic: Leaf is thin and spongy, capable of absorbing water.

Scaly: Thin scale like and colour less.

Crestaceous: Leaf is hard and brittle.

INFLORESCENCE

The mode of development and arrangement of flowers on the peduncle is called inflorescence.

Peduncle: The stalk bearing an inflorescence.

Pedicel: The stalk of the flower.

Sessile: Flowers without pedicel (stalk).

Rachilla: The axis of spikelet (Secondary axis of peduncle).

Racemose: The peduncle grows indefinitely and produces a number of flowers directly or indirectly in acropetal manner (older flowers at base and younger flowers at the top).

Simple raceme: Peduncle is simple, unbranched, elongated, producing many pedicellate, bracteate flowers in acropetally.

Spike: The peduncle is racemose, unbranched and produces bracteate sessile flowers acropetally.

Compound spike: The peduncle is branched and the branches produce bracteate sessile flowers acropetally. Those branches are called spikeletes.

Catkin: Peduncle is long, weak and drooping; on this peduncle develop many sessile, unisexual flowers acropetally.

Simple spadix: Peduncle is fleshy, unbranched and produces many sessile, bracteate, unisexual flowers acropetally; one of the bracts is modified, into a thick, leathery structure called spathe covering inflorescence.

Compound spadix: Peduncle is branched and on the branches develops many sessile, unisexual flowers acropetally.

Head/capitulum/anthodiusm: The inflorescence axis is condensed into a flattened disc. Many small sessile flowers (florets) are closely arranged on this receptacle. The entile inflorescence is covered and protected by involucre of bracts. If the inflorescence produces only one type of florets called homogamous head, if both the types of florets are produced, the ray florets arranged at the periphery and disc florets at the centre called heterogamous hed.

Simple corymb: The peduncle grows indefinitely producing many pedicellate flowers in an acropetal manner. The older flowers have longer pedicels and younger flowers have shorter pedicels. Thus all the flowers are brought more or less to the same height.

Compound corymb: Peduncle is branched and each branch is produced into a simple corymb.

Simple umbel: The peduncle is condensed and unbranched many pedicellate and bracteate flowers arise from its apex in cluster. All the flowers open in centripetal manner. At the base of flowers, all the bracts are united from a whorl called involucre.

Compound umbel: The peduncle is branched and each branch produces a simple umbel.

Cymose: The growth of peduncle is determinate stoped, due to the development of a flower at its apex. The inflorescence bears the older flowers at the apex and younger flowers at the base (basipetal arrangement).

Simple cyme/cymule: Inflorescence axis terminates into a flower, below this flower the main axis. Produces two branches which also terminate in to a flower.

Monochasial cyme: A cyme reduced to a single flower on each axis. (Only one branches arises).

Helecoid cyme: In monochasial cyme, the successive lateral branches develop only one side in a spiral manner.

Scorpioid cyme: In monochasial cyme, the lateral branches develop, successively on either side in a zigzag manner.

Dichasial cyme: After terminating into a flower the inflorescence axis produces two lateral branches below, these two branches repeat the same type of growth.

Poly chasial cyme: A determinate inflorescence having a whorl of lateral branches bearing younger flowers.

Glomerule: A very dense single cyme, distinguished from a head by blooming of the central flower first.

Thyrus: A panicle like cluster with main axis indeterminate and lateral axis determinate.

Cyanthium: The small, cup like specialized inflorescence bearing the floers on its base.

Verticellaster: A false whorl, composed of a pair of opposed cymes (Dichasial cyme and continues its growth as monochasial scorpioid cyme).

Hypanthodium: Fruit like inflorescence and bearing flowers on the inner wall of the cavity.

FLOWER

The flower is a modified shoot essentially for the reproduction of the plant.

Pedicel: Stalk of an individual flower.

Receptacle/torus: Enlarged or elongated end of the flower axis on which some or all of the flower parts are borne.

Bract: A more or less modified leaf subtending a flower or a flower cluster.

Bracteole: A secondary bract/a bractlet.

Bracteate: Flower with bracts.

Ebracteate: Flower without bracts.

Bracteolate: Flower with bracteole.

Involucre: A whorl of bracts subtending an inflorescence.

Involucel: A secondary involucres that subtends a part of an inflorescence.

Phyllary: A bract of the involucre.

Glumes: A pair of empty scale likes bracts at the base of a grass spikelet.

Lemma: A bract that usually encloses a flower in the spike let of grasses.

Palea: The inner bract usually smaller, subtending the grass flower in a spikelet.

Cupule: An involucre composed of bracts adherent by their base atleast.

Complete: A flower having all the four sets of floral organs i.e. sepals, petals, stamens and carpels.

Incomplete: A flower lacking one or more sets of floral organs.

Actinomorphic: With radial symmetry, the parts similar in size and shape, capable of division in any plane into equal halves.

Zygomorphic: The flower can be cut into two equal halves only in plane.

A symmetric flower: Flower cannot be cut into two equal halves.

Hermaphrodite/Bisexual: Flower with androecium and gynoecium.

Unisexual: A flower having only one sex.

Staminate: A unisexual flower with androecium.

Pistillate: A unisexual flower with gynoecium.

Neutral: Flower without fertile essential organs.

Monoecious: When male and female flowers are borne on same palnt.

Dioecious: When male and female flowers are borne on the different plant.

Poly gamous: If unisexual and bisexual flowers are borne on the same plant.

Poly gamomonecious: Some perfect and some staminate flowers are produced oon the same plant.

Androgamy: Having staminate flowers above and pistillate flowers below in the same spike.

Gynenandrous: Having pistillate flowers above and staminate flowers below in the same spike.

Dimerous: In each whorl the floral parts are two or in multiple of two.

Trimerous: The parts are 3's or multiples of three.

Teramerous: The parts are in 4's or multiple of five.

Pentamerous: The parts are in 5's or multiple of five.

Hypogynous: A flower is which all other floral organs are situated beneath the ovary (superior ovary).

Perigynous: A flower in which all other floral organs borne and arising from around the ovary (half superior).

Epigynous: The other floral parts are arranged along the margin of the thalamaus i.e. above the level of gynoecium (inferior ovary).

Acyclic/spiral: The floral parts are arranged spirally.

Cyclic: The floral parts are aarranged in circles.

Hemi cyclic: Perianth lobes are arranged in circles/whorls but stamens and carpels are arranged spirally.

Florets: Individual flowers in the inflorescence.

CALYX

Sepals: One of the separate part of a calyx, usually green and foliaceous.

Poly sepalous: Having sepals free.

Gamo sepalous: Sepals are fused.

Pappus: Sepals are reduced to hair like structures.

Calyculus: A stimulated calyx composed of bracts and bracteoles.

COROLLA

Petal: One of the separate parts of a corolla, usually coloured and showy.

Polypetalous: Having corolla of separate petals.

Gamopetalous: Having corolla of united petals.

Apetalous: Without petals.

Perianth: Calyx and corolla together constitute of the perianth.

Tepal: Individual member of perianth which is not clearly differentiated into calyx and corolla.

Achlamydeous: Without perianth.

Monochlamydeous: Perianth is arranged in one whorl.

Dichlamydeous: Perianth is arranged in one whorl.

Lodicule: Reduced perianth parts of a grass flower.

Aestivation: The mode of arrangement of perianth lobes during bud condition.

Open: Perianth lobes are free and are arranged with wide gaps between them.

Valvate: The perianth lobes are arranged closely with small gaps between them.

Twisted/convolute: Perianth lobes over lap each other i.e. one margin of the perianth lobe is covered by the next lobe.

Imbricate: In aestivation of five parts, one being exterior, one interior and rest three having one margin exterior and other margin interior (1+1+3 manner).

Ascendingly imbricate: Posterior petal is inner most.

Descendingly imbricate: Posterior petal is outer most.

Quincuntial: Perianth lobes over lap in 2+2+1 manner i.e. the margins of two perianth lobes are completely over lapped, and two other are external and the remaining one perianth lobe, one margin is inside and one out side.

Vexillary: When there are five petals, of which the posterior one is largest and it almost covers the two lateral petals, and the latter in their turn nearly overlap the two anterior or smallest petals.

COROLLA SHAPE

Cruciform: Petals four, arranged in a cross shaped manner each petal is clawed.

Caryophyllaceous: Petals five, clawed, each petal has a narrow stalk like base, with an apex arranged at right angles to the base.

Rosaceous: Petals five, each petal is narrow at base and broad, triangular at the apex. They are arranged symmetrically.

Papilionaceous: Petals five, zygomorphic corolla with one large posterior standard, two lateral wings and two inner most fused petals, the keel.

Coronary: At the tip of the petals, a crown like structure is formed with specialized cells or scales.

Tubular: The petals are fused to form a tubular structure.

Companulate: The gamo petalous corolla looks like a bell.

Infundibuliform/funnel shaped: The corolla lobes are fused to form a funnel shaped structure.

Rotate: The corolla has a short tube, at the apex of which the petals are arranged in right angles.

Hypocrateriform/salvar shaped: The corolla has a long tube; the petals are arranged on this tube at right angles.

Bilabiate: The fused petals form two lips like structure.

Ligulate: Strap shaped, three petals are fused to form a tongue shaped structure.

Personate: A bilabiate corolla having one large lip enclosing the other.

Gibbose: Swollen on one side, usually basally (corolla appendage).

ANDROECIUM

Androphore: A support or stalk on which the stamens are arised above the perianth.

Stamen: The unit of the androecium, consistsing of anther and filament.

Staminode: A sterile stamen.

Pollination: A mass of pollen grain adhering together and shed as a unit.

Filament: A stalk of a stamen.

Anther: The pollen containing part of a stamen.

Connective: The part of a stamen joining the two anther sacs.

Poly androus: With an indefinite number of stamens.

Didynamous: With four stamens, in two pairs of two different lengths.

Tetra dynamous: With six stamens, four longer than the outer two.

Inserted: Attached, as a stamen growing on the corolla.

Exerted stamens: The filaments of the stamens are longer than the petals, so they come out of the flower.

Adelphy: If the filaments of stamens in a flower are united leaving the anther lobes free.

Monodelphous: The filaments of all stamens are united to form only one bundle.

Dia delphous: The filaments of nine stamens are united into one bundle and the tenth posterior stamen remain free.

Poly delphous: The stamens are united to form many bundles by fusion of their filaments.

Syngenesious: The anther lobes are fused, leaving their filaments free.

Synandry: The filaments and anthers are fused completely.

Gynostegium: Stamens are attached to the stigmas of gynoecium.

Epipetalous: Stamens adhere (attached) to the corolla lobes.

Epipetalous/epiphyllous: Stamens adhere to the inner margins of the perianth lobes.

Epi sephalous: Stamens adhere to the sepals.

Gynandrous: Stamens adnate to the carpels as a whole or by their anthers only.

Haplo stamonous: All the stamens in a flower are arranged in one whorl.

Diplostemonous: All the stamens are arranged in two whorls outer whorl of the stamens alternate with petals.

Obdiplo stemonous: Stamens are arranged in two whorls and outer whorl is opposite to corolla lobes.

Dithecous: A two celled anther (two lobes).

Monothecous: Single lobe anther.

Introrse: An anther dehiscing towards the centre of the flower.

Extrorse: An anther dehiscing towards the periphery of the flower.

Latrose: An anther dehiscing laterally.

Adnate: Filament running the whole length of the anther from the base to the apex.

Basifixed: Filament attached to the base of the anther.

Dorsifixed: Filament attached to the back of the anther.

Versatile: Filament attached to the back of the anther at a point only so that the laterally can swing freely.

GYNOECIUM

The gynoecium is the collectively term for the female parts of a flower, the pistil or pishls (ovary, style and stigma).

Gynophore: The stalk on which an ovary is elevated above the floral axis.

Andro gynophore: An axis bearing both stamens and pistil above the point of perianth attachment, a gynophore associated with the androphore.

Pistillode: A rudimentary pistil present in some staminate flowers.

Ovary: Ovule bearing part of pistil.

Carpel: A mega sporophyll, a single, modified seed bearing leaf.

Mono carpellary: With one carpel.

Bicarpellary: With two carpels.

Tricarpellary: With three carpels.

Tetra carpellary: With four carples.

Penta carpellary: With five carpels.

Multi carpellary: With many carpels.

Syncarpous: Carpels are fused.

Superior: When a ovary is borne above the point of attachement of perianth and stamens.

Inferior: When a ovary is borne below attachement of all the floral parts.

Semi inferior/ semi superior: When the conditions is intermediate between superior and inferior.

Locule: Chamber or compartment of ovary.

Unilocular: With one chamber.

Bilocular: With two chambers.

Trilocular: With three chambers.

Tetralocular: With four chambers.

Pentalocular: With five chambers.

Multilocular: With many chambers.

Placenta: A place or part in the ovary where ovules are attached.

Placentation: The arrangement of ovules with in ovary.

Marginal placentation: The ovary is mono carpellary and unilocular. The placenta developes on the fused ventral margins of the carpel. The ovules are arranged on this placenta.

Axile placentation: The margins of the carpels fold inside and meet at the centre forming a central axis and many chambers. The ovules arise from this central axis which bears the placenta.

Parietal placentation: Ovary is unilocular and the ovules are attached to the place where the carpels are united. Number of placenta is equal to number of carpels.

Free central placentation: The ovules are borne on the central column with no septa present. Hence, the ovary is unilocular.

Basal placentation: Ovules develop from the base of the ovary.

Apical placentation: One or few ovules arising from the top of the ovary.

Super ficial: Ovules are borne not only on the ventral margins but on the inner surface of septa.

Style: A more or less elongated part of the carpel between the ovary and stigma.

Terminal: A style lying in the same straight line with the ovary.

Lateral: A style arising from the side of the ovary.

Gynobasic: A style arising from the depression in the centre of the ovary or directly from the thalamus.

Stylopodium: A disc like enlargement at the base of the style.

Stigma: The part of a carpel that receives the pollen.

Capitate: Like a globbose head.

Discoid: Disc shaped.

Plumose: Feather like.

Linear: Long and furrow.

Dumb bell shaped: Like a dumb bell.

Bifid: Divided in to two.

Sticky: A stigma producing a sticky liquid.

Gynostegium: A covering of the gynoecium of whatever nature.

Ovule: The egg containing unit of the ovary, after fertilization becomes the seed.

Ortho tropous: An ovule with a straight axis, micropyle at apex and hilum at base.

Anatropous: An ovule that is reversed, micropyle is close to the point of funiculus attachment.

Campylotropous: An ovule curved by un even growth so that its axis is approximately at right angles to its funicules.

FRUIT

The fruit is ripened ovary of the flower enclosing seeds.

True fruits: Fruits which are derived from fertilized ovaries.

False fruits: Fruits developed from other parts of flower such as calyx, thalamus and pedicel, other than ovary.

Multiple fruit: Fruits developed from all the flowers of inflorescence are associated closely and form a single composite fruit at maturity.

Aggregate fruit: Fruits are developed from a single flower having multicarpellary, apocarpous ovary. Each carpel developes into a fruit let. All these fruits are aggregated on a common pedicel to form a single aggregate fruit.

FLESHY FRUITS

Berry: A pulpy, indehiscent few to many seeded fruit derived from a gynoecium of fused carpels.

Pome: Fruit developed form bi or multicarpellary syncarpous inferior ovary surrounded by fleshy thalamus. Pericarp is hard.

Pepo: Fruit developed from tricarpellary, syncarpous unilocular, inferior ovary, the pericarp is differentiated into outer epicarp fused with a hard sheath like thalamus to form a rind fleshy mesocarp and smooth endocarp.

Hesperidium: Fruit developed from, multicarpellary, syncarpous multilocular, superior ovary. Pericarp is devided into leathery epicarp, having many glands, white papery mesocarp and the inner endocarp which has many chambers filled with juicy edible hairs.

Drupe: A fleshy, one seeded fruit with the seed enclosed in a stony endocarp.

DRY FRUITS

In the fruits, the pericarp remains dry and undifferentiated when ripe.

Dry dehiscent fruits: The fruits dehisce and liberate the seeds at maturity.

Follicle: A dry dehiscent fruit, formed from single carpel that splits on one side only.

Siliqua: The ovary is unilocular, but becomes bilocular due to the development of false septum called replum. At maturity the pericarp dehisces dorsiventrally from base to apex, leaving the seeds on the replum.

Silicula: A short and broad siliqua containing only a few seeds.

Capsule: Fruit at maturity dehisces by splitting along two or more lines of suture in various ways.

Pyxidium: A capsule with circum scissile dehisceence, the upper portion acting as a lid.

Dry indehiscent fruit: Fruits are normally one seeded and never dehisce even at maturity. The seeds are liberated out only after the disintegration of the pericarp.

Achene: A small, hard, dry, indehiscent fruit, one seeded in which the seed coat is free from the fruit wall.

Caryopsis: A small, hard, dry indehiscent fruit, in which the seed coat is fused with the fruit wall.

Cypsela: An achene with persistant pappus like calyx.

Nut: Fruit is characterized by stony pericarp, it encloses a single seed, the pericarp and the seed coat remain free.

Samara: It is winged dry indehiscent fruit.

Schizocarpic fruits: Dry fruits which show the characters of both the dehiscent as well as indehiscent fruit. At maturity the fruit splits into many one seeded bits called 'cocci' the seeds present in cocci are liberated only after the disintegration of the pericarp.

Lomentum: It is legume like, at maturity the fruit dehisces into one seeded bits (cocci) the fruit is constricted between the seeds.

Cremocarp: It is developed from bicarpellary ovary. In between the carpels a carpophore is present which an elongated thalamus is. At maturity the fruits dehisces into two one seeded mericarps attached to the carpophore.

Carcerulus: At maturity the fruit dehisces into four nutlets (bicarpellary syncarpous tetralocular ovary).

Regma: Fruit developes from tricarpellary, syncarpous trilocular, superior ovary, at maturity the fruits splits into three one seeded cocci.

Schizocarp: Fruit developes from multicarpellary, multi locular, syncarpous, superior ovary. At maturity it splits into many cocci.

Balausta: The fruit of pomegranate with a firm rind, berried within and crowned with the lobes of an adnote calyx.

SEED

The seed is a mature ovule, consisting of an embryo, with or without endosperm, and a surrounding protective coat.

Testa: The outer coat of the seed, usually hard and brittle.

Endosperm: Nutritive tissue surrounding and nourishing the embryo in seed plants.

Embryo: The rudimentary plant with in the seed.

Funiculus: Stalk of ovule, attaching it to placenta.

Hilum: Scar on seed coat, marking point of farmer attachment of seed to funicle.

Micropyle: The minute opening into an ovule for the entrance of the pollen tube, becoming a pit like mark on the mature seed.

Aril: An appendage or out growth from the hilum or funicular of a seed, often spongy and some time enveloping the seed.

Caruncle: An appendage at or about the hilum of some seed.

Cotyledons: An embryonic leaf of a seedling or in a seed.

Kernal: The nucellus of an ovule or of a seed i.e. the whole body within the seed coat.

Perisperm: Nutritive tissue surrounding embryo in some seeds derived from nucellus.

Raphae: The portion of the funiculus of an ovule, represented by a ridge.

Strophiole: An appendage of the hilum in some seeds.

❑❑❑

References

Aastveit, A.H. and Aastveit, K. 1990. Theory and application of open-pollination and polycross in forage grass breeding. *Theor. Appl. Genet.* 79: 618–624.

Abdurakhmonov, Ibrokhim. 2013. Role of Genomic Studies in Boosting Yield. *The icac recorder,* Vol. XXXI, No. 4, 2013.

Acquaah, G, 2002. Horticulture: Principles and Practices. New Jersey: Prentice Hall. ISBN 0130331252.

Acquaah, G. 2007. Principles of Plant Genetics and Breeding. Blackwell Eds., Malden, MA, USA, 569p.

Agrawal, R.L. 1998. Fundamentals of plant breeding and hybrid seed production. Science Publishers, Inc., Enfield, NH.

Ali, M. and Kumar, S., eds., 2006. Advances in Mungbean and Urdbean. Indian Institute of Pulses Research, Kanpur, India. 462 p.

Allard, R.W. 1999. Principles of Plant Breeding 2nd ed. John Wiley and Sons, NY, USA.

Allard, R.W. and Adams, J. 1969, Population studies in predominantly self-pollinating species. XIII. Intergenotypic competition and population structure in barley and wheat. *Am. Nat.* 103: 621-645.

Allen, A, 2008. "A Passion for Tomatoes". Smithsonian Institution.

Ames, IA. Tigchelaat, E.C., and V.W.D. Casali. 1976. Single seed descent: Applications and merits in breeding self-pollinated crops. *Acta Horticulturae* 63: 85–90.

Andrews, D.J. and Kumar, K. A. 1992. Pearl millet for food, feed and forage. *Advances in Agronomy* 48: 89–139.

Anil Grover and Deepak Pental, 2003. Breeding objectives and requirements for producing transgenics for major field crops of India. *Current science* 84(3).

Arora, R.K. 2000. Wild relatives of cultivated plants. *In Flora of India.* p. 218-234.

Arzu Manova, A.M. 1950. Biological characteristics of flowering and fruit setting in sesame. *Bull App. Bot. Genet. Plant Breed,* 28: 81-27.

B.S. Kurlovich and S.I. Repyev (eds.), St. Petersburg: 1995, the gene bank and breeding of grain legumes, *N. I. Vavilov Institute of Plant Industry,* 438p.

Baenziger, P.S., Russel, W.K., Graef, G.L. and Campbell, B.T. 2006. Improving Lives: 50 years of crop breeding, genetics, and cytology. *Crop Science.* 46: 2230–2244.

Baker, R.J. 1986. Selection Indices in Plant Breeding. *Cereal Research Communication,* Boca Raton, FL.

Balconi C, Lanzanova C, Motto M, 2010. Ribosome-inactivating proteins in cereals. In: Lord JM, Hartley MR (eds) Toxic plant proteins. *Springer, Heidelberg/London/New York* pp 149–166.

Bänziger, 2000. "Breeding for drought and nitrogen stress tolerance in maize: from theory to practice". *From Theory to Practice:* 7–9.

Baulcombe D, 2004. RNA silencing in plants. *Nature* 431: 356–363.

Belton, P.S. and J.R.N. Taylor. 2004, Sorghum and millets: protein sources for Africa.

Bernardo, R. 2002. Breeding for quantitative traits in plants. *Stemma press,* Woodburry MN.

Berrie, A. 1977. An introduction to the botany of the major crop plants. Heyden and son, London.

Bioversity International, 2006. Crop wild relatives. *Bioversity International,* Rome.

Bliss FA, 2007. Education and preparation of plant breeders for careers in global crop improvement. *Crop Science* 47: S250–S261

Blum, A. 2006. Drought adaptation in cereal crops: a prologue. In: Drought Adaptation in Cereals. pp. 3–15. Ribaut, J. M., Ed., the Harworth Press, Inc., New York.

Bosland, P.W, 1998. "Capsicums: Innovative uses of an ancient crop". In J. Janick. Progress in new crops. Arlington, VA: ASHS Press. pp 479–487.

Brar, D. S. and Khush, G. S. 2003. Utilization of wild species of genus Oryza in rice improvement. In: Nanda, J.S. and Sharma, S. D. (eds.), Monograph on Genus *Oryza,* pp 283-309.

Bravo A, Gill SS, Soberon M, 2007. Mode of action of Bacillus thuringiensis toxins and their potential for insect control. *Toxicon* 49: 423–435.

Brown J, Caligari PDS and Campos HA, 2014. Plant Breeding, 2nd edn. Chichester: Wiley Blackwell.

Busbice, T.H. 1970. Predicting yield of synthetic varieties. *Crop Science.* 10: 265–269.

Chahal, G.S and Gosal, S.S., 2002. Principles and procedures of Plant Breeding, Alpha Science International, United Kingdom.

Chang, T.T. 1976. The origin, evolution, cultivation, dissemination and diversification of Asian and African rices. *Euphytica,* 25(1): 425-441.

Chapman and Hall. Hill J, Becker HC and Tigerstedt PMA, 1998. Quantitative and Ecological Aspects of Plant Breeding. London: Chapman and Hall.

Chickpea: An ancient crop for the modern *world*http://exploreit.icrisat.org/page/chickpea/685/60. ICRISAT.

Cleveland, D.A., D. Soleri, and S.E. Smith. 2000. A biological framework for understanding farmers' plant breeding. *Economic Botany* 54: 377-394.

Collard BCY, Mackill DJ, 2008. Marker assisted selection: an approach for precision plant breeding in the twenty-first century. *Philos Trans R Soc Lond Ser B Biol Science* 363: 557–572.

Comstock, R. E. 1979. Use of inbred lines in reciprocal recurrent selection. *Crop Science.* 19: 881–886.

Comstock, R.E., H.F. Robinson, and P.H. Harvey. 1949. A breeding procedure designed to make maximum use of both general and specific combining ability. *Agronomy Journal* 41: 360–367.

Cooper M., O.S. Smith, G. Graham, L. Arthur, L. Feng, D.W. Podlich, 2004. Genomics, genetics and plant breeding: A private sector perspective. *Crop Science.* 44: 1907-1913.

Crotalaria juncea. *USDA NRCS Plant Guide.*

Dahlia; *et al.* 2009. "Consumer Preference for Indigenous Vegetables". World Agroforestry Centre.

Damodaran T., Hegde D.M. 2007. Oilseeds situation: a statistical compendium. Directorate of Oilseed Research, Hyderabad.

Dangl JL and Jones JDG, 2001. Plant pathogens and integrated defense responses to infection. *Nature* 441: 826–833.

Das, A. P. 2000. Perspective of plant biodiversity. Bishen Singh Mahendra Pal Singh, Debra Dun, India.

De Maagd RA, Bravo A and Crickmore N, 2001. How Bacillus thuringiensis has evolved specific toxins to colonie the insect world. *Trends Genet* 17: 193–199.

Directorate of Rice Research. 2007. DRR Perspective Plan: Vision 2025, published by Directorate of Rice Research, Hyderabad.

Dodds PN, Lawrence GJ and Catanzariti A-M, The T, Wang C-IA, Ayliffe MA, Kobe B, Ellis JG, 2006. Direct protein interaction underlies gene-for-gene speci fi city and coevolution of the flax resistance genes and fl ax rust avirulence genes. *Proc Natl Acad Science* U S A 103: 8888–8893

Donald, C.M, 1968. The breeding of crop ideotypes. *Euphytica* 17(3): 385–403.

Doraiswamy, I., A. Basu, and K. P. Chellamani, 1998. "Development of Fine Quality Jute Fibers". Colourage. Nov. 6–8, 2p.

Dorothea Bedigian, 2010. Sesame: The genus Sesamum. CRC Press. ISBN 978-0-8493-3538-9.

Durner J, Shah J, Klessig DF, 1997. Salicylic acid and disease rsistance in plants. *Trends Plant Science* 2: 266–274.

Duvick DN, 2001. Biotechnology in the 1930s: the development of hybrid maize. *Nat Rev Genet* 2: 69–74.

E.A. Wernsman and D.F. Matzinger, 1980. Tobacco. American Society of Agronomy- Crop science society of America.

Ehlers, J. D. and Hall, A. E., 1997. Cowpea (*Vigna unguiculata* L. Walp.). *Field Crops Res.*, 53: 187-204.

Ellis J, Dodds P and Pryor T, 2000. Structure, function and evolution of plant disease resistance genes. *Curr Opin Plant Biol* 3: 278–284.

Falconer, D.S., 1989. Introduction to Quantitative Genetics. 3rd Ed. Longman. Burnt Mill.

FAO, 2008. Establishment of a global network for the in situ conservation of crop wild relatives: status and needs. *FAO*, Rome.

FAOSTAT. Crop production. 2015 Available from: *http:// faostat3.fao.org/ home/E.*

Ferh, W.R. 1987. Principles of cultivar development. *McMillan*, New York.

Flor HH, 1971. Current status of the gene-for-gene concept. Annu Rev Phytopathol 9: 275–296.

Food Science & Technology, 15(2): p. 94–98.

Forster BP, Till BJ, Ghanim AMA, Huynh HOA, Burstmayr H and Caligari PDS, 2015. Accelerated plant breeding. CAB Reviews No. 43: 1749-8848.

Frankel, R and Galun, E, 1977. Pollination mechanisms, Reproduction and Plant Breeding, Springer-Verlag, New York.

Frey, K. J., Browning, J. A., and Simons, M. D, 1975. Multiline cultivars of autogamous crop plants. *SABRAO Journal*. 7: 113–123.

Friedman AR and Baker BJ, 2007. The evolution of resistance genes in multi-protein plant resitance systems. *Curr Opin Genet Dev* 17: 493–499.

Frisch, M. and Melchinger, A.E., 2005. Selection Theory for Marker-assisted Backcrossing. Genetics: Published Articles Ahead of Print, published on March 31, 2005 as *10.1534/genetics.104.035451*

Fu, T. D., 2000: Breeding and Utilization of Rapeseed Hybrid (second edition). Wuhan: Hubei.

Gallais, A. 2003. Quantitative genetics and breeding methods in autopolyploid plants. *INRA*, Paris, France.

Gao, Ji-Ping, et al. 2007. "Understanding Abiotic Stress Tolerance Mechanisms: Recent Studies on Stress Response in Rice." *Journal of Integrative Plant Biology* 49(6): 742-750.

Gepts P, 2002. A comparison between crop domestication, classical plant breeding, and genetic engineering. *Crop Science* 42: 1780–1790.

Gepts P and Hancock J, 2006. The future of plant breeding. *Crop Science* 46: 1630–1634.

Goldbach R, Bucher E and Prins M, 2003. Resistance mechanisms to plant viruses: an overview. *Virus Res* 92(2): 207–212.

Greyson, R.I. 1994. The Development of Flowers. New York/Oxford: Oxford University Press.

Hajjar, R. and Hodgkin, T., 2007. The use of wild relatives in crop improvement: a survey of developments over the last 20 years. *Euphytica*, 156: 1-13.

Havey, M. J. 2004. Application of genomic technologies to crop plants: opportunities and challenges. *Crop Sci.* 44: 1893–1895.

Hayward MD, Bosemark NO and Romagosa I (eds.) 1993. Plant Breeding: Principles and Prospects. London.

Hegde DM, 2009. Vegetable oils scenario: approaches to meet the growing demands. Indian Society of Oilseeds Research, Directorate of Oilseed Research, Hyderabad.

House, L.R., Verme, B.N., Ejeta G., Rana, B.S., Kapran, I., Obilana, A.B. and Reddy, B.S.V. 1997. Developing countries breeding and potential of hybrid sorghum. In 'Proceedings of the International Conference on Genetic Improvement of Sorghum and Pearl Millet". 23-27 Sept. 1996. *Lubbock, TX, USA. INTSORMIL Publication* 97-5. pp 84-96.

Hulbert SH, Webb CA, Smith SM and Sun Q, 2001. Resistance gene complexes: evolution and utilization. *Annu Rev Phytopathol* 39: 285–312.

Indira, M. and Kurup, P.A., 2013. "Black Gram: A Hypolipidemic Pulse". *Natural Product Radiance.* 2(5).

Jensen, N.F. 1978. Composite breeding methods and the DSM system in cereals. *Crop Science.* 18: 622–626.

Jinks, J. L., and Pooni, H. S, 1984. Comparison of inbred lines produced by single seed descent and pedigree inbreeding. *Heredity* 53: 299–308.

Jones JDG and Dangl JL, 2006. The plant immune system. *Nature* 444: 323–329.

Kale D.M. and Mouli, C. 1984. Hybridization technique in groundnut. *Ind. J. of. Genetic. Plant. Breed.* 44: 379-384.

Khush, G. S. 1999. Green revolution: preparing for the 21st century. *Genome* 42: 646–655.

Kooyers, N.J., 2015. The evolution of drought escape and avoidance in natural herbaceous populations. *Plant Science.* 234: p. 155–162.

Kumar K.A. and Andrews D.J. 1993. Genetics of qualitative traits of pearl mullet: A review. *Crop Science* 33: 1-20.

Legumes Of The World Royal Botanic Gardens, Kew". *www.kew.org.* Retrieved 2015.

Legumes. *J. Wiley and Sons,* Chichester, UK, pp 411-427.

Leins, P. and Erbar, C. 2010. Flower and Fruit. Stuttgart: Schweizerbart Science Publishers.

Martin, Franklin W., 1982. "Okra, Potential Multiple-Purpose Crop for the Temperate Zones and Tropics". *Economic Botany*. 36(3): 340–345.

Matzinger, D.F., and Wernsman, E. A, 1980. Population improvement in self-pollinated crops. In "World Soybean Research Conference II" (F. T. Corbin, ed.). Westview Press. Boulder, CO.

Maxwell FG and Jennings PR, 1980. Breeding plants resistant to insects. Wiley, New York.

McDowell JM and Simon SA, 2006. Recent insights into R gene evolution. *Mol Plant Pathol*: 7: 437–448.

McGregor, S.E. 1976. Insect pollination of cultivated crop plants. *U.S. Dep. Agric. Handb.* 496. Gov. Print. Office, Washington, D.C.

Millan, Teresa; Heather, J.Clarke; Kadambot and H.M.Siddique; et.al, 2006. "Chickpea molecular breeding: New tools and concepts". *Euphytica*. 147: 81-103.

Mishra, P.K. 1985. Heterosis breeding and exploitation of male sterility in mustard. PhD thesis, P.G. School IARI, New Delhi.

Murphy, D.J. 2002. Biotechnology and the improvement of oil crops—genes, dreams and realities. *Phytochem. Rev.* 1: 67–77.

Noel Kingsbury, 2009. Hybrid: The History and Science of Plant Breeding. University of Chicago Press. p. 140.

Painter RH, 1958. Insect resistance in crop plants. University Press of Kansas, Lawrence.

Pan, N. C., A. Day, and K. K. Mahalanabis, 2000. "Properties of Jute". *Indian Textile Journal*. 110: 5.

Parker, M.A. 1991, Nonadaptive evolution of disease resistance in an annual legume. *Evolution* 45: 1209-1217.

Parlevliet J.E. 1979. Components of resistance that reduce the rate of epidemic development. *Annu Rev Phytopathol* 17: 203–222.

Passioura, J.B. 1996. Drought and drought tolerance. *Plant Growth Regul.* 20: 79–83.

Patil, Shreekant S. 2014. Conventional Breeding of Cotton Needs to Change. *The icac recorder,* Vol. XXXII, No. 3: 2014.

Peng, S., Khush, G.S., Virk, P., Tang, Q. and Yingbin, Z, 2008. Progress in Ideotype Breeding to Increase Rice Yield Potential. *Field Crops Research*, 108: 32-38.

Peter Griffee, 2000. "Saccharum Officinarum". Food and Agriculture Organization of the United Nations.

Poehlman J.M. and Sleper D.A., 1995. Breeding Field Crops. Ames, IA: Iowa State University Press.

Posselt, U.K. 2000. Constraints in the selection of parents for synthetic cultivars. In: Proceedings of 23rd fodder crops and amenity grasses section meeting of Eucarpia, Oct 2000, Azores, Portugal, pp 1–4.

Posselt, U.K. 2003a. Recombination and progeny testing in recurrent selection. *Vortr Pflanzenzüchtg* 59: 110–115.

Posselt, U.K. 2003b. Heterosis in grasses. *Czech J. Genet. Plant. Breed.* 39: 48–53.

Putnam, D.H.; Oplinger, E.S.; Teynor, T.M.; Oelke, E.A.; Kelling, K.A.; Doll, J.D. , 1991. "Peanut". Alternative Field Crops Manual, NewCROP Center, Purdue University.

Putt, E.D., 1997. "Early history of sunflower". In A.A. Schneiter. Sunflower Technology and Production. Agronomy Series. 35. Madison, Wisconsin: *American Society of Agronomy*. pp 1–19.

Raghav Ram; David Catlin; Juan Romero and Craig Cowley, 1990. "Sesame: New Approaches for Crop Improvement". Purdue University.

Rai, B. 1970. Heterosis Breeding. Agro-Biological publication Azd Nagar, New Delhi, India pp 183.

Raven, P.H. and G.B. Johnson. 1991. Understanding biology. Mosby Yearbook, Inc., St, Louis.

Redden R. J., and Jensen, N. F, 1974. Mass selection and mating systems in cereals. *Crop Science*. 14: 345–350.

S.K. Gupta, 2005. Practical Plant Breeding, Agrobios, India.

Sampath, S. 1962. The genus Oryza: Its taxonomy and species relationships. *Oryza* 1: 1-29.

Sarkarung, S., Singh, O.N., Roy, J.K., Vanavichit, A. and Bhekasut, P, 1995. Breeding Strategies for Rainfed Lowland Ecosystem. *Proc. Int. Rice Res. Conf.,* Manila, IRRI, 709-720.

Sattler, R. 1973. Organogenesis of Flowers. A Photographic Text-Atlas. University of Toronto Press.

Saxena, K.B. 2004. Pigeonpea. In: Plant Breeding – Theory and Techniques, Gupta, S.K. *Agrobios Publishers,* Jodhpur. Niles, J.A. and Feaster, C.V.

1984. Cotton Breeding In: Cotton (ed) Kohel, R.J. and Lewis, C.I. Agronomy Monograph. 24th American Soc. Med., Wisconsin.

Sharpe, Peter, 1998. "Sugar Cane: Past and Present". Southern Illinois University.

Shewry, Peter R, 2009, "Wheat", *Journal of Experimental Botany*, 60(6): 1537–1553.

Shull GH, 1909. A pure line method in corn breeding. *American Breeders Association Report* 5: 51–59.

Simmonds, N.W. and Smartt, J. 1999. Principles of Crop Improvement, 2nd Ed. Blackwell Science, Oxford, UK.

Simmonds, N.W. 1976. Evolution of crop plants. Longman, London and New York.

Simmonds, N.W. 1995, The relation between yield and protein in cereal grain. *J. Sci. Food Agric.* 67: 309-315.

Singh, B.; Ajeigbe, H.A.; Tarawali, S.A.; Fernandez-Rivera and S.; Abubakar, M., 2003. "Improving the production and utilization of cowpea as food and fodder". *Field Crops Research.* 84: 169–150.

Singh, G., Sekhon, H.S. and Kolar, J.S., eds., 2005. Pulses. Agrotech Publishing Academy, Udaipur, India, pp. 59-94, 485-518.

Singh, K.B. and Saxena, M.C., eds., 1993. Breeding for Stress Tolerance in Cool Season Food.

Singh. B. D. 2004. Plant Breeding: Principles and methods (ed), Kalyani Publishers.

Snape, J.W., and Riggs, T.J. 1975. Genetical consequences of single seed descent in the breeding of self-pollinating crops. *Heredity* 35: 211–219.

Snape, J.W. and Simpson, E. 1984. Early generation selection and rapid generation advancement methods in autogamous crops. *In* "Efficiency in Plant Breeding," *Proc. 10th Congr. EUCARPIA* Pudoc, Wageningen, The Netherlands.

Soumendra chakraborty and Tapash dasgupta, 2011. Principles and plant breeding methods of field crops in India. New Delhi Publishers.

Tandon, J.P. and Sharma, S.P. 2002. Guidelines for crop variety testing under ALL-India Coordinated Crop Improvement Projects. Indian Council of Agricultural Research, New Delhi.

Technical Information Section. 1998. Breeding for Yield Improvement Needs Changes. the *icac recorder*, Vol. XVI, No. 1: 1998.

Tester, Mark; Langridge and Peter, 2010. "Breeding technologies to increase crop production in a changing world". Science. *AAAS: American Association for the Advancement of Science.* 327(5967): 818–822.

The Biology of *Gossypium hirsutum* L. and *Gossypium barbadense* L. (cotton). *ogtr.gov.au*

Van der Maeson, L.J.G. 1995. "Pigeonpea Cajanus cajan", pp. 251–5 in Smartt, J. and Simmonds, N. W. (eds.), *Evolution of Crop Plants.* Essex: Longman.

Van der Plank J.E. 1982. Disease resistance in plants. *Academic,* New York.

Vaughan, D.A. and Lu, B; Tomooka, N, 2008. "The evolving story of rice evolution". Plant Science. 174(4): 394–408.

Vavilov, N.I. 1926. Studies in the origin of cultivated plants. Institute of Applied Botany and Plant Breeding, Leningrad.

Wang, W., Vinocur, B. and Altman, A. 2007. "Plant responses to drought, salinity and extreme temperatures towards genetic engineering for stress tolerance." *Planta* 218: 1-14.

Weberling, Focko, 1992. Morphology of Flowers and Inflorescences. Cambridge: Cambridge University Press. ISBN 9780521438322.

Yadav, S.S., Redden, R., Chen, W. and Sharma, B., eds., 2007. Chickpea Breeding and Management. CAB International, Oxfordshire, U.K., pp. 1-13, 193-212, 233-245, 391-416.

Zhu JK. 2002. Salt and drought stress signal transduction in plants. *Annu. Rev. Plant Biol.* 53: 247-273.

□□□